TIMBER ENGINEERING

PRACTICAL DESIGN STUDIES

TIMBER ENGINEERING

PRACTICAL DESIGN STUDIES

E. N. Carmichael CEng MIStructE FFB AIWSc
Consulting Engineer

London New York
E. & F. N. Spon

First published 1984 by
E. & F. N. Spon Ltd
11 New Fetter Lane, London EC4P 4EE
Published in the USA by
E. & F. N. Spon
733 Third Avenue, New York NY10017

Typeset by Keyset Composition, Colchester
Printed in Great Britain by J. W. Arrowsmith, Ltd, Bristol

ISBN 0 419 12690 2 (cased)
ISBN 0 419 12700 3 (paperback)

British Library Cataloguing in Publication Data

Carmichael, E. N.
Timber engineering.
1. Timber—Handbooks, manuals, etc.
2. Building, Wooden—Handbooks, manuals, etc.
I. Title
624.1′84 TH1101

ISBN 0-419-12690-2
ISBN 0-419-12700-3 Pbk

Library of Congress Cataloging in Publication Data

Carmichael, E. N. (Ernest Norman)
Timber engineering.

Bibliography: p.
Includes index.
1. Building, Wooden. 2. Timber. I. Title.
TA666.C345 1984 694′.2 83-14679
ISBN 0-419-12690-2
ISBN 0-419-12700-3 (pbk.)

Contents

Preface

Sadly, the state-of-the-art of structural engineering still favours instruction in the more traditional building materials such as concrete, steel and brickwork, whilst paying little, if no, attention to timber. It is my contention that timber should be given more time and attention when preparing the syllabus for structural engineering studies. Contrary to the belief of many, it is not an inferior subject and when studied well it provides a stimulating field in which to pursue one's career.

All good designers cultivate a flexibility of thought and a tolerant approach to the acceptance of theoretical solutions by others, particularly in regard to their translation into the physical state. The transference from paper to practice is all too often lost in a highly theoretical solution to a problem. In the use of timber, perhaps more than any other material, the building site application must always be foremost in the mind of the designer when applying his theory. A tolerance of around ±5% to the arithmetical answer is considered reasonable when designing in timber.

The majority of structural timber used in the United Kingdom is to be found in single occupancy housing, flats, maisonettes, light industrial buildings, small office and institutional developments and those diverse buildings used in the agricultural industry. In addition it is unusual for timber to be used to form structures more than four storeys in height. (The Scottish regulations limit its use to three storeys.)

Because of the large variability of the material we are dealing with and the factors of safety employed (see Chapter 3 on stress grading) it is not necessary to be highly theoretically accurate in the approach to most designs, though of course this should be no excuse for a lax approach to a problem. For example, the method of obtaining maximum bending moments in non-symmetrically loaded simply supported beams by breaking down the load patterns and summating the maximum centre span moments for these patterns from standard tables, although not theoretically accurate, is acceptable for most purposes. Therefore, it is not the intention of this book to dwell on the theory behind the standard solutions offered to the numerous examples given but merely to offer a path to a sound solution satisfying both theory and practice.

Notwithstanding this, I would add that it would be unwise for anyone to administer timber design without an understanding of the theory of structures and use of materials. In short a designer must have a sound training in all levels of structural design and theory.

Casting my mind back to my student days, I recall how often I found that theoretical solutions were divorced from a practical application. What I have attempted to do in this book is first to provide an introduction in the broadest possible sense to the world of timber and its structural use, including as much information on materials as is required to develop the more commonly designed elements. Secondly, I offer step-by-step solutions to the various structural elements which when put together form the framework of the majority of timber buildings provided in the United Kingdom today. In addition, I have set tasks taken from everyday design office occurrences which I hope the student will attempt to solve without first referring to the solutions which I offer.

The reader will find that those tasks and all worked examples are based upon the recommendations of CP 112:Part 2:1973, which is currently under review and will eventually be published as BS 5268:Part 2 ‘Permissible stress design’. However, like most new issues it will be some time before this standard is fully understood and integrated into the system and there will of course be a period of grace when both the old and the new will be permitted to be used. Rather than delay publication I feel that the correct procedure is to allow the natural course of events to take place and to review my book at some later date when the new standard is well established.

I have assumed that the reader is suitably skilled in the important subject of Theory of Structures and does not require me to comment other than to draw attention to the relevance of some terms which are common to timber alone.

Finally, I suggest an approach to drafting calculations which I consider will lead to clearly-defined thinking and loss of ambiguity; an attribute to which all designers should strive.

I hope that this book will provide the background and impetus for a more advanced study of the art of timber engineering to some, if not all, who choose to read it.

Acknowledgements

When a book of this nature is undertaken it is not possible to achieve its completion without the assistance and co-operation of a good many individuals and organizations. In this respect I am, in particular, most grateful to Margaret Cocks, Ingrid Tomlinson and Stephanie Rooke for their many hours typing from what, at times, were my almost indecipherable notes and sketches. I am thankful to my wife and family for their patience and encouragement throughout many evenings and weekends when those numerous chores remained unattended.

I wish also to thank certain organizations who have kindly consented to the use of references to work which is subject to their particular copyright. These include extracts from CP 112:Part 2:1971, reproduced by permission of the British Standards Institution, 2 Park Street, London W1A 2BS, from whom complete copies of the standard can be obtained; the Building Research Establishment to whom Crown copyright is acknowledged in adapting Figs 2.1 and 3.1; the Council of Forest Industries of British Columbia and Hydro-Air International (UK) Limited.

Finally, I acknowledge with much affection the lively encouragement given to me by my father-in-law, Olaf Bugge, who always expressed so much interest in my work and to whom I had hoped to present a copy of this book. Sadly his death came before its completion and he is greatly missed.

E.N.C.
May 1983

Glossary of terms

The following terms are those which are used in the text throughout this book. There are many more but as we are dealing with only a limited number of subjects, reference is confined to those which are necessary.

Bending

f_{apar}	applied bending stress parallel to grain
f_{gpar}	grade bending stress parallel to grain
f_{ppar}	permissible bending stress parallel to grain
M and M_o	applied bending moment
$\bar{M}$	moment capacity of the section

Deflection

d	deflection due to bending
d_p	permissible deflection
d_v	shear deflection
E	modulus of elasticity
E_{mean}	arithmetical mean value E
$E_{min.}$	statistical minimum value of E for one member acting alone
E_N	statistical minimum value of E for N pieces acting together
N	number of sections acting together to support a common load

Compression

C_{apar}	applied compression stress parallel to grain
C_{gpar}	grade compression stress parallel to grain
C_{ppar}	permissible compression stress parallel to grain
C_{aperp}	applied compression stress perpendicular to grain
C_{gperp}	grade compression stress perpendicular to grain
C_{pperp}	permissible compression stress perpendicular to grain

Shear

V	applied shear
v	applied horizontal shear stress
$\bar{V}$	shear capacity of section
v_g	grade horizontal shear stress
v_p	permissible horizontal shear stress
v_{pa}	panel shear stress
V_r	rolling shear

Tension

t_{apar}	applied tensile stress parallel to grain
t_{gpar}	grade tensile stress parallel to grain
t_{ppar}	permissible tensile stress parallel to grain

Section properties

A	total cross-sectional area
A_b	bearing area
b	overall thickness of section
D	depth of bearing
d	overall depth (width) of section
F	form factor
G	modulus of rigidity
I	second moment of area
I_x	second moment of area about major axis
I_y	second moment of area about minor axis
L	actual length of column
l	effective length of column
L_c	clear span of beam
L_e	effective span of beam ($L_c + D$)
l/r	slenderness ratio
Q	first moment of area
Q_x	first moment of area about major axis
Q_y	first moment of area about minor axis
r	radius of gyration
r_x	radius of gyration about major axis
r_y	radius of gyration about minor axis
Z	section modulus
Z_x	section modulus about major axis
Z_y	section modulus about minor axis

Wind loading coefficients

(CP 3, Chapter V, Part 2: 1973 'Wind')

C_f	force coefficient
C_p	pressure coefficient
C_{pe}	external pressure coefficient
C_{pi}	internal pressure coefficient
q	dynamic wind pressure
q_d	design wind pressure
S_1	topography factor
S_2	ground roughness, building size and height above ground factor
S_3	a statistical factor relating to the intended life of the building
V	basic (map) wind speed
V_s	design wind speed

Modification factors

(CP 112: Part 2: 1971 'The structural use of timber')

K_1, K_2, etc.	load increase or reduction factors as described in the text

Load sharing increase factor is constant at 1.1

Abbreviations

BCLMA	British Columbia Lumber Manufacturers Association, Templar House, 81 High Holborn, London WC1.
BRE	Building Research Establishment, Garston, Watford, Herts.
BSI	British Standards Institution, 2 Park Street, London W1A 2BS.
BWPA	British Wood Preservative Association, 150 Southampton Row, London WC1.
COFI	Council of Forest Industries of British Columbia, Templar House, 81 High Holborn, London WC1.
CPA	Chipboard Promotional Association, 7a Church Street, Esher, Surrey.
FPDA	Finnish Plywood Development Association, Finland House, 56 Haymarket, London SW1.
FPRL	Forest Products Research Laboratory, Princes Risborough, Aylesbury, Bucks. (Now known as the Princes Risborough Laboratory – PRL.)
HMSO	Her Majesty's Stationery Office, York House, Kingsway, London WC1.
TRADA	Timber Research and Development Association, Hughenden Valley, High Wycombe, Bucks.

1

Timber materials

1.1 Softwoods (of coniferous trees)

Most softwoods used in the United Kingdom are imported and the use of home grown material in the structural field is small. We obtain our softwoods from the USSR, Poland, Sweden, Finland, Norway, Canada, and, to a much lesser degree, from the USA. There are very many different species of softwood and their strength and properties vary enormously, being dependent upon such factors as region of growth, rate of growth, moisture content, specific gravity and temperature. Most of the more commonly used softwoods are referred to in CP 112:Part 2:1971 'The structural use of timber' and their standard, common and botanical names are reproduced in Table 1.1. In addition to these, we have those timbers specifically of Canadian origin and these are listed in Table 1.2.

Whilst examining this particular list, it should be noted that one of the more commonly imported species is the kiln-dried Spruce-Pine-Fir surfaced and graded to Canadian lumber standards. Commercially, this is referred to as CLS (S-P-F) and is used in growing quantities, particularly in the field of timber frame housing.

The principal species included in this group are White Spruce, Engelmann Spruce, Lodgepole Pine and Alpine Fir. The dominant species in the group is White Spruce and, consequently, in shipping discussions it is frequently referred to by this particular species' name. Of all of these softwoods, those most commonly used in the construction industry are European Redwood, European Whitewood and Canadian Spruce-Pine-Fir and it is to these three species that we will generally direct our attention in design.

A most important factor, which must not be overlooked when selecting the softwood, is the question of moisture content whilst in service and hence durability. It is well known that timber rapidly loses strength with any increase in moisture content up to the saturation level of 25% known as the 'green' state, but, above this percentage, strength is not significantly affected. Conversely, strength will increase with a drop in moisture down to 18% known as the 'dry' state.

Table 1.1 Names and densities of some structural timbers

Standard name	*Botanical species*	*Other common names*	*Approximate density at a moisture content of 18% (kg/m³)*
SOFTWOODS			
(a) Imported			
Douglas Fir	*Pseudotsuga menziesii*	BC Pine	590
Western Hemlock (unmixed)	*Tsuga heterophylla*	BS Hemlock	540
*Western Hemlock (commercial)	*Tsuga heterophylla*	Hembal	530
Parana Pine	*Araucaria angustifolia*	—	560
Pitch Pine	*Pinus palustris* *P. elliottii* *P. caribaea*	Longleaf Pitch Pine Southern Yellow Pine Nicaraguan Pitch Pine Honduras Pitch Pine	720
Redwood	*Pinus sylvestris*	Baltic Redwood	540
Whitewood	*Picea abies* *Abies alba*	Baltic Whitewood European Whitewood	510
Canadian Spruce	*Picea glauca* *Picea sitchensis*	Western White Spruce Eastern Canadian Spruce Sitka Spruce	450
Western Red Cedar	*Thuja plicata*	B.C. Red Cedar	3[illegible]
(b) Home-grown			
Douglas Fir	*Pseudotsuga menziesii*	—	560
European or Japanese Larch	*Larix decidua* *Larix leptolepsis*	Larch	560
Scots Pine	*Pinus sylvestris*	Scots Fir	540
European Spruce	*Picea abies*	Norway Spruce	380
Sitka Spruce	*Picea sitchensis*	—	400
HARDWOODS			
(a) Imported			
Abura	*Mitragyna ciliata*	Subaha	590
African Mahogany	*Khaya* spp.	Khaya	590
Afrormosia	*Pericopsis elata*	Kokrodua	720
Greenheart	*Ocotea rodiaei*	—	1060
Gurjun/keruing	*Dipterocarpus* spp.	—	720
Iroko	*Chlorophora excelsa*	Mvule	690
Jarrah	*Eucalyptus marginata*	—	910
Karri	*Eucalyptus diversicolor*	—	930
Opepe	*Nauclea diderrichii*	Kusia	780
Red Meranti/red Seraya	*Shorea* spp.	—	540
Sapele	*Entandrophragma cylindricum*	—	690
Teak	*Tectona grandis*	—	720
(b) Home-grown			
European Ash	*Fraxinus excelsior*	—	720
European Beech	*Fagus sylvatica*	—	720
European Oak	*Quercus robur*	—	720

Not all species of timber are equally available at all times. There are many other hardwoods besides those quoted but, since peculiarities may exist in particular timbers, advice should be sought from the appropriate authorities as to the suitability of alternative species and the stresses to be used.

* May include Western Balsam (*Abies* spp.) and some Mountain Hemlock (*Tsuga mertensiana* Carr.).

Table 1.2 Names and densities of Canadian timbers and species combinations

Standard name	*Botanical species*	*Other common names*	*Approximate density at a moisture content of 18% (kg/m³)*
Douglas Fir-Larch	*Pseudotsuga menziesii* *Larix occidentalis*	BC Pine Western Larch	590
Hem-Fir	*Tsuga heterophylla* *Tsuga mertensiana* Carr *Abies amabilis* *Abies grandis*	Western Hemlock Mountain Hemlock Amabilis Fir Grand Fir	530
Princess Spruce*	*Picea glauca* *Picea rubens* *Picea mariana* *Abies balsamea* *Pinus banksiana*	White Spruce Red Spruce Black Spruce Balsam Fir Jack Pine	450
Western White Spruce*	*Picea glauca* *Picea engelmanni* *Picea mariana* *Pinus contorta* *Pinus banksiana* *Abies lasiocarpa* *Abies balsamea*	White Spruce Engleman Spruce Black Spruce Lodgepole Pine Jack Pine Alpine Fir Balsam Fir	450

* Because of similarity of strength properties these two species groups are collectively designated as Spruce-Pine-Fir.

Therefore, when selecting softwood, if strength and durability is a long term consideration, a species will have to be chosen which is known for these qualities. For example, when constructing envelopes to swimming pools, where high humidity and various chemical elements exist, Douglas Fir is very often used because of its durability performance. It would, of course, be vacuum pressure treated but, nevertheless, its own properties make it an ideal material for this purpose.

Availability is another important factor which must not be overlooked. Each year the exporting countries make their 'offer' prices to our importing agents and, therefore, it is not uncommon for the emphasis of usage to change depending upon those offer prices. One year the price may favour CLS(S-P-F), the next year could well see a switch to Russian Whitewood, and so on. The designer, therefore, must be ever-ready to advise accordingly if it is felt that the softwood required will impose a premium on the project. Again, for example, the architect may require a specific surface finish which cannot be obtained by the use of the more common softwoods available and, therefore, a cost penalty will almost certainly apply, perhaps without the architect being aware and to which he would probably expect guidance from the timber designer.

The American Redwood is another species which is imported in small quantities to this country. In its native country it is, of course, extensively used and is known for its durability, colour, strength, lightness in weight, straight grain and knot free appearance. Any student of timber engineering should, if given the opportunity, visit the Redwood forests of California and see the sobering image of a plant which can grow in excess of 90 m in virtually a straight vertical line, have a diameter of around 6.0 m and reach ages in excess of three thousand years.

To summarize, the selection of softwood is influenced by such factors as strength requirement, durability in service and availability. However, generally speaking, the softwoods most commonly used in our construction industry for structural purposes are European Redwood, European Whitewood and Canadian Spruce-Pine-Fir.

1.2 Hardwoods (of deciduous trees)

The use of hardwoods in the construction industry is not extensive and is normally limited to elements where large loads need to be considered such as those encountered in the spacing of portal frames or where geometric restraints exist, like limited depth above a lintel opening.

Many years ago, this country had its own bountiful supply of the British Oak and examples of its extensive use in Tudor times exist to this day. Many an old house, public house and hotel boasts traditional oak beams and posts with the axe mark of the tradesman of the day. Today, the British Oak is still used but is harder to come by in its seasoned state.

In the main, we employ imports to satisfy our needs and we primarily look to South America, the African continent and Malaysia for our supplies. Again, the more commonly used hardwoods are given in CP 112:Part 2:1971 'The structural use of timber' and their standard common and botanical names are once again listed in Table 1.1. The majority of imported hardwoods are used in joinery industries for such items as domestic furniture, kitchen cabinets, windows and doors, etc., where factories are tooled for the rigorous use of hardwood. Hardwoods are also used extensively in the ship and boat building industries.

The use of hardwoods to satisfy building construction needs should, until such time as the industry desires it, be limited. This statement is made for the simple reason that in most manufacturing units and on the site, the tools used are geared to softwood and hardwood plays havoc with saw blades, planes, drills, etc. In addition, some dense hardwoods can make site fixings extremely difficult to achieve, all adding to production and labour costs. So the designer must beware the specification of hardwood and the limitations that it imposes upon the construction industry. Where hardwood becomes a necessity, the designer should always consider and consult with the factory (if any) for tooling, the builder for practicalities on site usage, and the supplier for ease of availability of species.

The commonest hardwood encountered in the structural field is Keruing. It is a hardwood of reasonable workability and cost and will accept most site fixing media. Unfortunately, it does have a tendency to shake badly and this should be noted when considering its end use condition.

Where hardwoods are to be made a feature, then of course Keruing would not be considered and one would seek a species which gives those obvious qualities required of feature work: good textured grain; small knots; low susceptibility to shaking, and amenability to good finishing. Some alternatives to hardwoods are flitch construction, ply box sections, glue-laminated construction and built up sections. All of these do, of course, make use of softwood but carry the penalty of manufacturing costs which should be measured against the specification of solid rectangular hardwood sections.

In summarizing this section, it may be said that the building designer should use hardwoods sparingly, specify with care and consult with all concerned.

1.3 Plywood

The use of plywood in construction is extensive and this material has a permanent place in timber design. It can be put to many uses where strength and rigidity are required – stressed skin panels, ply web and ply box beams, wall units, floor and roof sheathing, truss gusset plates, temporary formwork, templates, packaging – these are only a few examples but they serve to indicate the material's broad application and versatility. Plywood is a manufactured product and has been with us for around 100 years. It is produced from thin layers of softwood or hardwood or a combination of both, glued together generally in alternating horizontal and vertical layers to give varying overall thicknesses depending upon the number of layers used. There is theoretically no limit to the overall thickness but, in practice, the commonest thicknesses used are those indicated in Table 1.3. Some may be tongued and grooved. Normally, there is an odd number of plies to achieve a balanced construction because the outer layers are invariably of the same species with the grain running in the same direction, i.e. parallel to the width or parallel to the length of the sheet. The net result is a building material which, depending upon the grading chosen, can be extremely strong, rigid and durable. The layers are called 'veneers' and the number used in the overall product are sometimes referred to as the number of 'layups' (see Fig. 1.1).

The introduction of synthetic resin adhesives has made it possible to eliminate the problem of delamination extending the life of the plywood to that of the wood used and rarely is plywood specified in the construction field which does not possess a weather- and boil-proof quality resin adhesive (WBP). Other qualities are boil-resistant (BR) good weather-resistant

Table 1.3

Thickness (mm)	*t and g*	*Sizes and species*
8		*Sizes*
9		Generally plywood is sold in imperial sizes with the most popular
9.5		being:
12.0	√	(i) 2440 × 1220
12.5	√	(ii) 3050 × 1220
16.0	√	*Species*
19.0	√	Most popular species are:
25.0	√	(i) Canadian Douglas Fir
25.5	√	(ii) Finnish Birch throughout
		(iii) Finnish Birch faced with softwood cores

t and g = tongued and grooved.

√ = available.

qualities but not for long-term exposure, moisture-resistant (MR) and interior (INT) with only limited weather resistance.

The commonest plywoods used in the United Kingdom are the Douglas Fir and Birch faced plywoods with the Douglas Fir plywood being extensively imported from Canada. However, we do have a strong manufacturing industry of our own whose growth expanded rapidly during the Second World War owing, at this time, to the restriction on imports from Europe and America.

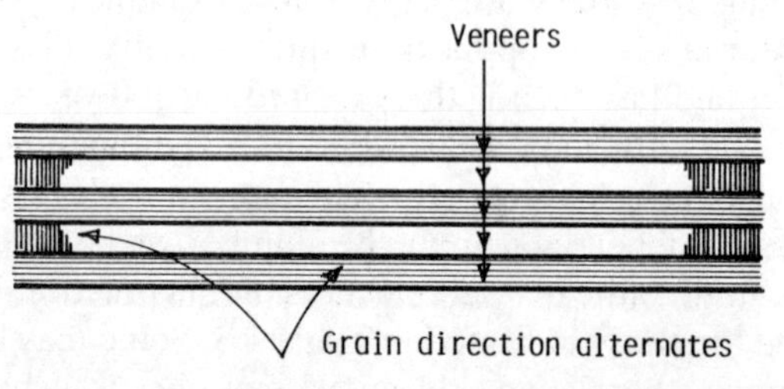

Fig. 1.1

In order that we may put the plywood to structural use, we need to know the basic permissible stress values which can be adopted. Most of these values are expressed in relation to the direction of the face grain relative to the direction of the load, i.e. face grain parallel to the load or face grain perpendicular to the load (Fig. 1.2).

Since plywood has a very wide variety of constructions, thicknesses and species, it is impossible to obtain the mechanical properties from test data alone. Therefore the stress values given in CP 112:Part 2:1971 'The structural use of timber' are obtained from mathematical calculations derived by examining basic test data carried out on 3-ply construction.

As the testing and grading of plywood is a subject in itself, it is beyond the scope of this book to cover it in its entirety and so the student is strongly recommended to examine the publications mentioned for further reading at the end of this book. There are, however, two types of stress which are intrinsic to plywood and which, therefore, require additional explanation. These are the terms 'panel shear' and 'rolling shear'.

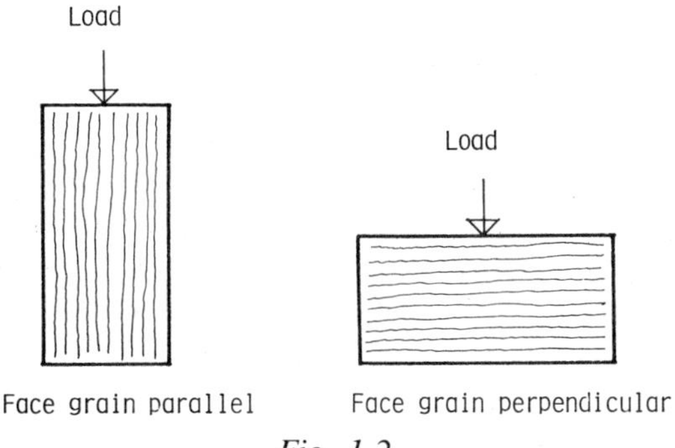

Fig. 1.2

1.3.1 Panel shear

This shear is sometimes referred to as 'shear through the thickness' simply because when considering the intensity of the design shear stress, the whole thickness of the plywood is taken into account. The result is the effective shear strength of the plywood as a whole. This strength may be increased by rotating the face grain such that it is 45° to the direction of the shear load. In practice, this is only occasionally done because of the wasteful cutting patterns induced in the sheets. It is more common to increase the plywood thickness or to reinforce locally, where possible, as in box beam construction at the supports.

1.3.2 Rolling shear

This stress occurs in concentrated local areas where the plywood is joined to other members and it is induced by bending moments and transverse loads. For example where, in a stressed floor panel the plywood sheathing passes over the supporting floor joist or where in I-sections or box beam sections the plywood is connected to solid timber chords (Fig. 1.3).

The term 'rolling shear' is derived from the mode of failure under test where failure occurs in those veneers having grain direction perpendicular to the applied load and the fibres appear to have rolled off one another. This phenomenon can occur either within the gross section between veneers (Fig. 1.4) or at the junction of the outer face of the plywood with the joined

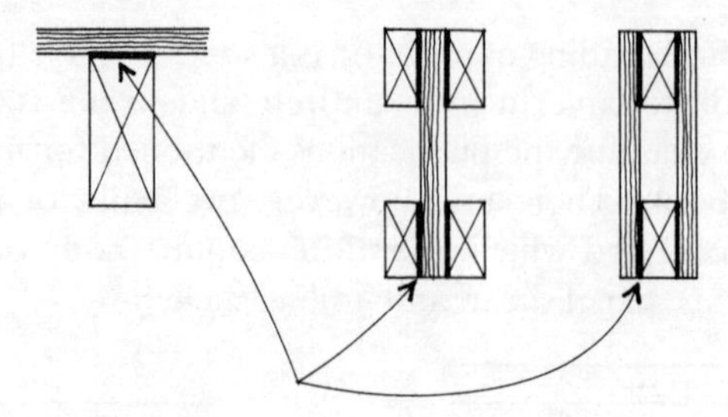

Fig. 1.3

member (Fig. 1.3) provided that, in the latter case, the face grains are perpendicular to one another. If the face grains are parallel, then the glue will permit the full resistance of the joined woods to develop. As permissible rolling shear stress is small in magnitude, it forms a very important feature in the design, particularly as a 50% reduction factor is employed in the code against the permissible design stress.

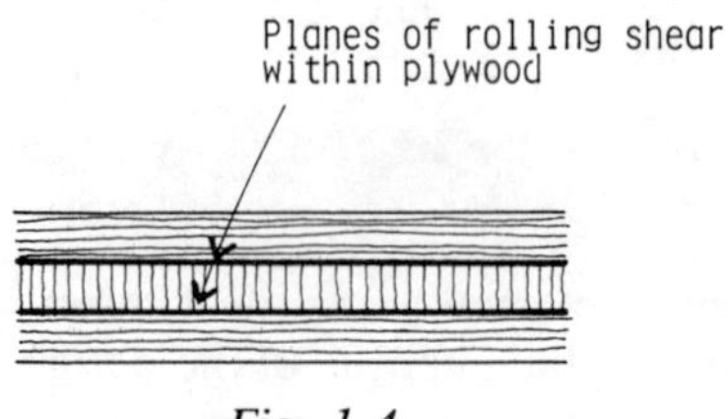

Fig. 1.4

1.3.3 Summary

To summarize, plywood is a manufactured product obtained from the use of many different softwood and hardwood species. Synthetic glues allow the full development of the wood veneer strengths. It is strong, rigid and durable, being produced in several quality grades, thereby making for a versatile and popular building material.

1.4 Blockboards, particleboards and fibreboards

1.4.1 Blockboards

Blockboards and laminboards are composite panels collectively referred to as core plywoods. The blockboard product is a central core of sawn timber battens, almost square in section, covered by an outer layer of plywood to each face giving 3-ply board. Alternatively, an additional outer layer may be provided giving 5-ply board (Fig. 1.5).

The board is normally available in thicknesses between 12 mm and 25 mm with core strips between 7 mm and 30 mm wide. The outer veneers are of plywood and the face grain of the outer layer varies in disposition according to the ply layup. In the 3-ply board, it is always perpendicular to the core grain direction whilst in the 5-ply the inner veneer is once again perpendicular to the core ply and the outer veneer may be either parallel with or perpendicular to

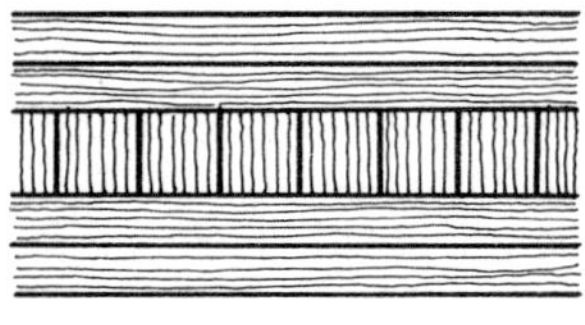

3-ply blockboard

Fig. 1.5

the inner veneer. In practice, it is normally parallel to the inner veneer. The construction allows for full bonding between veneers and core strips but the core strips are seldom bonded one to the other.

The laminboards are much the same except for the width of the strips forming the core. These are narrower and vary from 1.5 mm to 7 mm and they are glued together to form the core (Fig. 1.6). Generally speaking, laminboard is denser than blockboard because of the amount of glue used, and, in

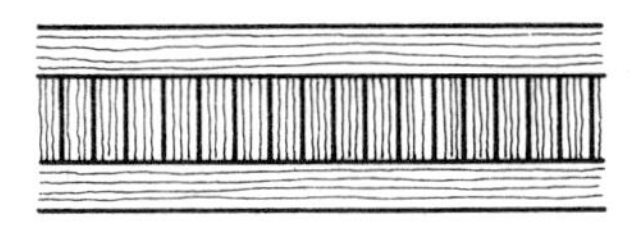

Fig. 1.6

addition, rotary peeling allows the use of denser species than those used in blockboard.

Although blockboards and laminboards are readily available, they are rarely used for structural purposes and are more commonly associated with interior non-loadbearing work. Their stress values are not covered in the current code of practice for timber and so it is essential for the designer to discuss the product with the manufacturer if contemplating use in a structural sense.

1.4.2 Particleboards

Of the particleboards, the commonest used for structural purposes is known as 'chipboard'. It varies in strength and density according to the amount of chips and resin used in the manufacture. Thicknesses vary from 9 mm through to 25 mm and, like plywood, thicknesses of 12 mm and above may be obtained with tongued and grooved edges. Again, sizes can be either imperial or metric, i.e. 2440 × 1220 or 2400 × 1200.

The greatest use for chipboard, in the construction field, is for flooring, and its thickness is dependent upon the chosen spacing of the supporting members and the load to be carried. Great care has to be taken with its on-site use because normally chipboard is not moisture resistant and it will rapidly deteriorate on wetting. Persistent wetting will cause its eventual failure and so the specification of its protective finish must be carefully considered and executed. Some products available are specified as moisture resistant but this only implies very short term; essentially while the erection process is taking place.

Another important consideration in design has to do with the method of fixing. Round wire nails should never be used where variability and frequency of loading is a factor. Because of the coarse fibrous content, round wire nails do not have an adequate grip and nail 'popping' is commonplace. Therefore, fixings should either be by means of annular ring nails or screws, especially developed for fixing into chipboard.

There is much literature supplied by the manufacturers of chipboard which relates to density, floor grading quality, frequency of fixings, etc. and the designer must always refer to this literature when using chipboard for structural purposes.

1.4.3 Fibreboards

Fibreboards as structural materials, particularly in the field of stressed skin claddings, are a growing market. However, because there are so many different products available and because research tends to be specific to a product, a definitive discussion is not possible within the scope of this book. Suffice to say that dense, strong and to some extent durable fibreboard products are available and the student can become more aware of their properties and field of usage by contacting the manufacturers for more detailed information. Detailed reference should also be made to the recent additions to CP 112:Part 2:1971, where their structural use is covered.

2

Some physical properties and their effects on strength

2.1 Introduction

In the past, a great deal of our knowledge of the correct use of various species of timber was born of many years of experience rather than a detailed knowledge of the natural properties of wood; availability was also a prime factor. Contrary to this situation, it is now readily accepted that a much more detailed knowledge of the properties of timber is required in order that we may determine which species best suit the construction industry and which give greater economy in design. This is particularly important in forestry management where selection for the regeneration of the natural product is of paramount importance. The possible exploitation of unusual species is another consideration. When using timber for construction purposes, it is essential to know its strength properties, such as bending, tension, compression, etc. In order to arrive at these values, test procedures have to be adopted which take into account the enormous variability of these properties, not only between species but within samples of the same species. Standard laboratory test procedures on small clear specimens have, therefore, been derived which arrive at the strength properties and from which working stress values may be determined. The code of practice which governs these test procedures is British Standard 373:1957 'Methods of testing small clear specimens of timber'.

There are two important physical properties of wood which have an effect upon its strength and hence its selection as a building material. These are moisture content and specific gravity. We will discuss each in turn to obtain a broad understanding, as distinct from a detailed one, of their significance in the field of timber engineering design. In addition, we will comment on the physical property of rate of growth.

2.2 Moisture content

Moisture content is the weight of water in the wood expressed as a percentage of the weight of the oven-dry wood. It is determined by weighing the sample prior to and immediately following drying and expressing the loss of weight as a percentage of the oven-dry weight. Drying is carried out in laboratory controlled conditions at a constant temperature of 103 ± 2°C until the weight becomes constant. In its natural state, wood contains a considerable amount of moisture held partly in the cell walls and partly in the cell cavities. During seasoning, most of the water in the cell cavities is lost, leaving a condition known as the 'fibre-saturation point'. With the continuation of controlled seasoning, the cell walls also lose their moisture and the timber shrinks, gaining strength in the process. Above the fibre-saturation point, changes in

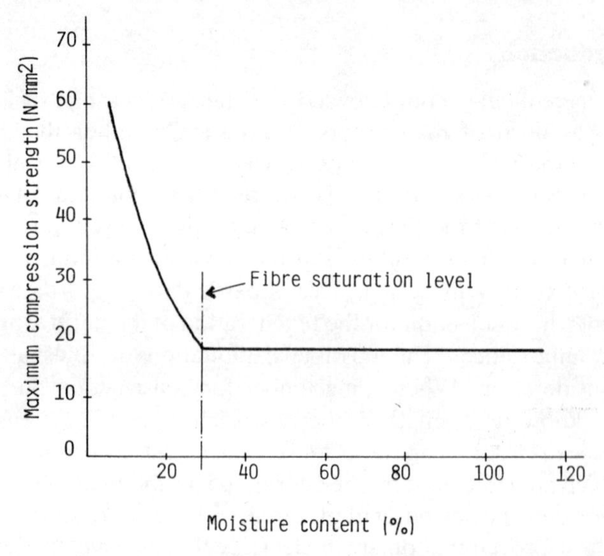

Fig. 2.1 The relation of compression strength to moisture content (Scots Pine)

the moisture content have no significant effect on the strength of wood. Fig. 2.1 shows how this phenomenon may be clarified by plotting the results of compression tests on a particular species against varying percentages of moisture. It will be seen from this graph that the fibre-saturation point occurs at around 25–30% and 25% is generally accepted as being a norm in timber strength assessment.

The trend of change of strength with change of moisture content is similar for most strength properties but the magnitude of the change varies from one property to another. The change in compression strength, for example, is more than that for bending strength which, in turn, is more than that found in modulus of elasticity. Adjustments of strength values for percentages below the fibre-saturation point are found by employing a mathematical equation derived by extensive research and study in the early 'thirties. It is, however, beyond the scope of this broad outline to go further into this and the student is recommended to the further reading given at the end of this book.

2.3 Nominal specific gravity

Specific gravity or relative density is the ratio of the weight of a substance to that of an equal volume of water. Density is the weight per unit volume normally expressed as kg/m^3. Specific gravity in timber research is indefinable because the weight of wood in a given volume changes with the shrinkage and swelling which follows a change in moisture content. So, in testing of wood, a nominal specific gravity is determined based on the volume of wood at the time of test and its weight when oven-dried. Specific gravity varies considerably, not only between species but within individual pieces of the same species and its value must be found by using standard deviation methods and plotting a Gaussian curve (see Chapter 3 on stress grading). Once the specific gravity is known, the various strength tests on moisture content are adjusted to take into account its effect upon these strength values.

2.4 Rate of growth

In past years, in the field of research, interest had been shown in using the rate of growth of trees as a measure of strength. However, it has been shown conclusively that rate of growth has little measurable effect upon the strength properties of wood and so is of no significant value.

This does not mean, however, that rate of growth serves no useful purpose; quite the contrary. Trees from the more northerly latitudes have a slower rate of growth and, consequently, produce smaller knots which all makes for a better quality. Commercially, there is therefore considerable interest in the rate of growth in regard to the end usage, for example, in joinery where small knots and, predominantly, clear timber are desirable.

3

Stress grading

3.1 Introduction

The study and application of stress grading in timber technology is absolutely essential if one is to develop an understanding of, and indeed a sympathetic feel for, timber as a building material.

Before an engineer can apply his skills to a given problem he must be given the physical and mechanical properties of the material and timber is no exception – rule of thumb, so often used in the past, no longer applies in a subject where our understanding of the material has grown so rapidly in recent times.

There are now two ways of stress grading timber; one is visual and the other is mechanical. We will go on to examine both and to see what effect they have on the end product.

3.2 Visual grading

3.2.1 Strength

Before the techniques of visual stress grading can be applied the strength of timber must be determined. It is a fact that the properties of timber vary considerably, not only between species but also within species and so in order to assign strength values proper laboratory investigations are made on small clear specimens 20 mm × 20 mm in size taken from a wide range of species. The requirements for strength testing in the United Kingdom involve six strength tests and two physical tests on each specimen and these are

(a) static bending
(b) impact bending
(c) compression parallel to grain
(d) hardness
(e) shear parallel to grain
(f) cleavage.

Included are the physical properties of moisture content and specific gravity. It is beyond the scope of this book to enter into the detail involved and the student is once again recommended to refer to the further reading suggested at the end of this book.

It should be noted that these strength tests are based on 'green' or fibre saturated timber which is generally accepted as being around 25%. This is because it is established that moisture content has a very important influence on strength, the higher the moisture content the lower the strength. However, above the fibre-saturation level there is no significant reduction in strength. Therefore by testing in the green state consistent values can be anticipated at this moisture level which may be accurately adjusted for the dry state of around 18% moisture content.

3.2.2 Basic stresses

Once the strength values are known the first stage is to derive the basic stress from these test values, which is in effect a safe stress for an ideal structural member free from all strength-reducing characteristics. The second stage is the assessment of the influence of defects on strength and this is taken into account in the formulation of grading rules.

The diversity in the various properties is found by drawing a 'normal' or Gaussian curve based on the arithmetic mean and the standard deviation of a number of test results. An example of this curve is shown in Fig. 3.1 for the property of compression.

From the histogram build up to the curve it is readily observed that the smaller the number of tests the larger the variability of results. As the number of tests increases so the variability decreases to give a more accurate result and a pattern begins to emerge. To this pattern may be fitted an arithmetically calculable curve – the Gaussian curve as previously described.

The properties of this curve are well known and they may be used to forecast the proportion of results which will be greater or less than a given value. It is known from the theory of the normal curve that 98% of the values lie within the range: mean $\pm 2.33 \times$ standard deviation; the remaining 2% lie outside this range. By applying a factor of safety to these results the basic stresses may be determined. Generally this factor of safety is taken as 2.25 for most properties in timber. Thus the basic stresses for each species and property are obtained from the results of standard tests on small clear specimens, in the green state, by dividing the statistical minimum value by the appropriate safety factor for that property.

The properties usually required are bending, compression parallel and perpendicular to the grain, tension, horizontal shear and modulus of elasticity.

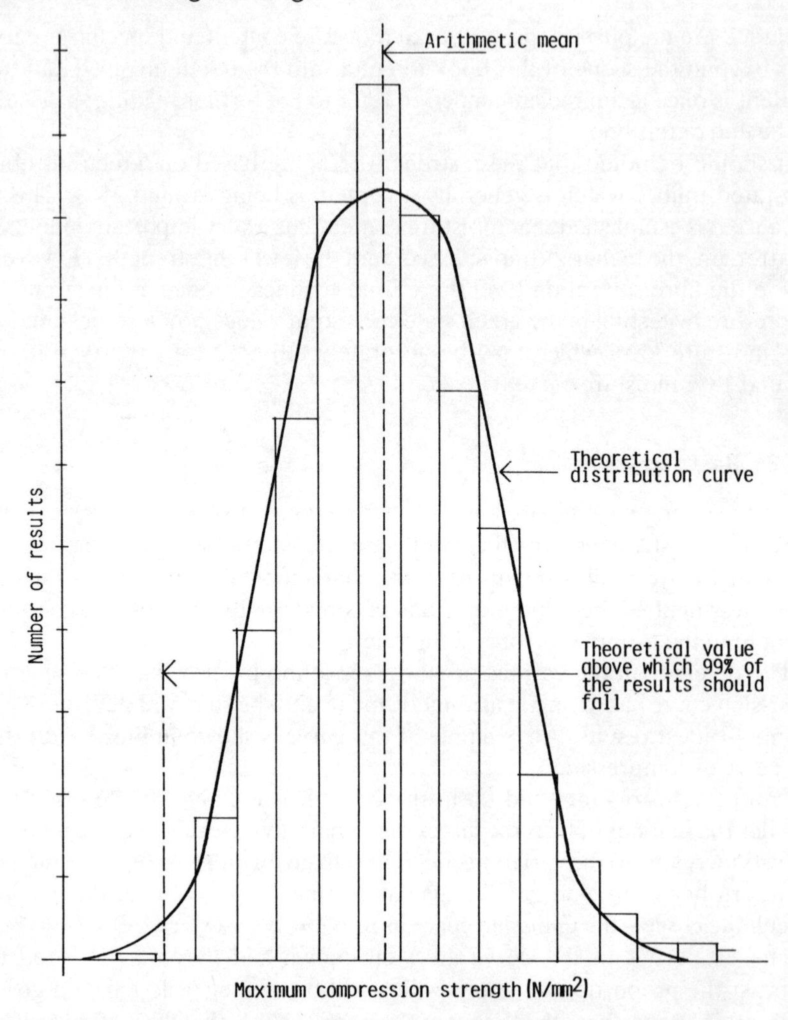

Fig. 3.1 Frequency distribution of maximum compression strength

3.2.3 Natural defects

As stated earlier the second stage is to examine the influence of the various natural defects on the strength and thereby arrive at a stress value which is acceptable for these defects.

The natural defects which must be considered are listed as follows with further explanatory notes by the side of each item:

Defect	*Notes*
Knots 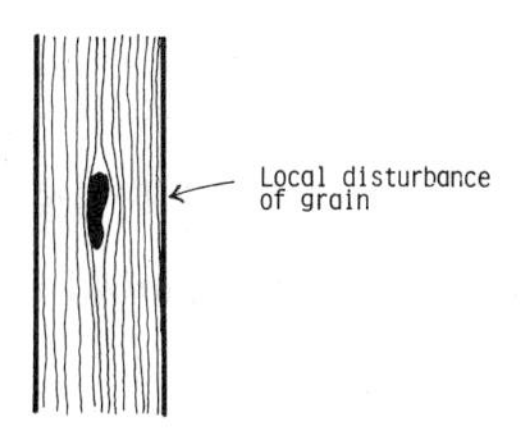	The weakening effect of a knot is brought about by the local disturbance in the grain direction it produces and it is not due to any inferiority in the material of the knot.
Wane 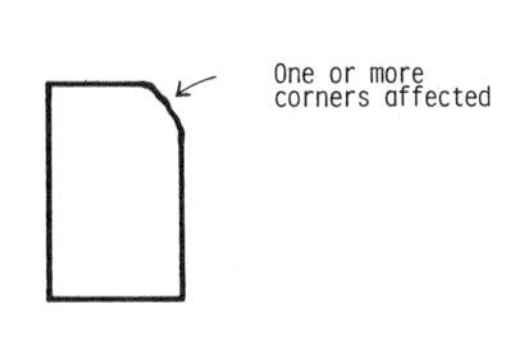	This is a reduction in the cross-sectional area of the rectangular timber section across the corners due to the section being taken from a location close to the outer circumference of the tree.
Slope of grain	This is the measure of the deviation of the fibres from the longitudinal axis of the piece. If fibres occur at an angle then any force applied along the longitudinal axis will create components of force on those fibres thus reducing strength. Timber is much weaker across the grain than along the grain and so excessively cross-grained timber is undesirable.
Rate of growth 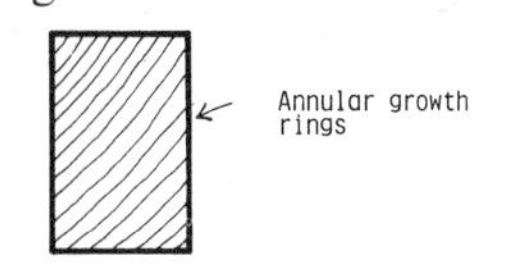	Although this is not as important as other features a limitation is imposed. Indicated by the average number of growth rings per 25 mm.
Fissures (resin pockets similar) 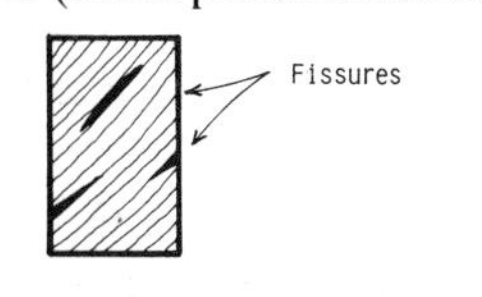	A fissure is any separation of fibres in a longitudinal plane and includes checks, shakes and splits. Their existence reduces the cross-sectional area resisting shear and bending stresses.
Bow, spring, twist, cup (for definitive diagrams refer to BS 4978)	These defects do not affect the strength of timber and grading rules are generally for obvious visual and practical reasons.

Sapstain (Blue stain)	This is not a structural defect and is limited only by visual acceptance.
Wormholes	Permitted to a slight extent provided that there is no active infestation. Wood wasp holes are not permitted.
Decay	Decayed wood should not be accepted.

3.2.4 Codes of Practice

There are currently two British Standard Codes of Practice which may be referred to for visual stress grading and these are CP 112:Part 2:1971 'The structural use of timber' (Appendix A) and BS 4978:1973 'Specification for timber grades for structural use'. There are fundamental differences between these codes which are briefly explained in the following sections.

3.2.5 CP 112:Part 2:1971

Appendix A to this code gives basic grading rules for visual assessment. They are formulated by fixing various strength ratios and specifying limitations for the previously listed defects such that they give the required ratio. Four ratios are fixed at 40%, 50%, 65% and 75% and the design stresses are applied accordingly. These stress grades are commonly known as the 'numerical grades' and they are now recognized to be inaccurate in comparison to the visual and machine grading rules of BS 4978.

They are however currently still acceptable and commonly used in the field of visual stress grading. This is because there are no rules related to marking and the stress grader does not have to prove competence by examination. Therefore, it permits the experience of those long established in the handling of timber to apply their skills in determining the appropriate numerical grade. Similarly, an experienced timber designer may also, if required, grade from this code without further training. The design stresses given for the various species and the basic numerical grades are reproduced in Tables 3.1 and 3.2 for green and dry conditions, respectively.

In addition to these stress grades the code allows for the grouping of species which is a convenient means of allowing flexibility of choice when clear availability is not known. Unfortunately such grouping does lead to a reduction in permissible stress values but nevertheless it does have practical merit. Because hardwoods cover a much larger range it is not considered feasible to group them.

Table 3.3 gives three strength groups for softwoods and Tables 3.4 and 3.5 indicate the respective green and dry stress levels.

Table 3.1 Green stresses and moduli of elasticity

Standard name	*Bending and tension parallel to the grain*					*Compression parallel to the grain*					*Compression perpendicular to the grain*			*Shear parallel to the grain*					*Modulus of elasticity for all grades*	
	Basic (N/mm^2)	*75 Grade (N/mm^2)*	*65 Grade (N/mm^2)*	*50 Grade (N/mm^2)*	*40 Grade (N/mm^2)*	*Basic (N/mm^2)*	*75 Grade (N/mm^2)*	*65 Grade (N/mm^2)*	*50 Grade (N/mm^2)*	*40 Grade (N/mm^2)*	*Basic (N/mm^2)*	*75/65 Grade (N/mm^2)*	*50/40 Grades (N/mm^2)*	*Basic (N/mm^2)*	*75 Grade (N/mm^2)*	*65 Grade (N/mm^2)*	*50 Grade (N/mm^2)*	*40 Grade (N/mm^2)*	*Mean (N/mm^2)*	*Minimum (N/mm^2)*
SOFTWOODS																				
(a) Imported																				
Douglas Fir	15.2	11.4	9.7	7.6	5.9	11.0	8.3	6.9	5.5	4.5	1.79	1.59	1.31	1.72	1.24	1.10	0.83	0.69	10 300	5 900
Western Hemlock (unmixed)	13.1	9.7	8.3	6.6	5.2	10.3	7.6	6.6	5.2	4.1	1.38	1.17	1.03	1.52	1.10	0.97	0.76	0.62	9 000	5 500
Western Hemlock (commercial)	11.7	8.6	7.6	5.9	4.5	9.0	6.6	5.9	4.5	3.4	1.38	1.17	1.03	1.38	1.03	0.90	0.69	0.55	8 600	5 200
Parana Pine	11.7	8.6	7.6	5.9	4.5	10.3	7.6	6.6	5.2	4.1	1.52	1.31	1.10	1.52	1.10	0.97	0.76	0.62	8 300	4 500
Pitch Pine	15.2	11.4	9.7	7.6	5.9	11.0	8.3	6.9	5.5	4.5	1.79	1.59	1.31	1.72	1.24	1.10	0.83	0.69	10 300	5 900
Redwood	11.7	8.6	7.6	5.9	4.5	8.3	6.2	5.2	4.1	3.1	1.52	1.31	1.10	1.38	1.03	0.90	0.69	0.55	7 600	4 100
Whitewood	11.7	8.6	7.6	5.9	4.5	8.3	6.2	5.2	4.1	3.1	1.38	1.17	1.03	1.38	1.03	0.90	0.69	0.55	6 900	4 100
Canadian Spruce	11.0	8.3	6.9	5.5	4.5	8.3	6.2	5.2	4.1	3.1	1.38	1.17	1.03	1.38	1.03	0.90	0.69	0.55	8 300	5 200
Western Red Cedar	9.0	6.6	5.9	4.5	3.4	6.2	4.5	3.8	3.1	2.4	1.03	0.90	0.76	1.24	0.90	0.76	0.62	0.48	6 200	3 800
(b) Home-grown																				
Douglas Fir	14.5	10.7	9.3	7.2	5.9	10.3	7.6	6.6	5.2	4.1	1.72	1.52	1.31	1.38	1.03	0.90	0.69	0.55	9 000	4 500
Larch	13.8	10.3	9.0	6.9	5.5	9.7	7.2	6.2	4.8	3.8	1.79	1.59	1.31	1.52	1.10	0.97	0.76	0.62	9 000	4 500
Scots Pine	11.0	8.3	6.9	5.5	4.5	8.3	6.2	5.2	4.1	3.1	1.72	1.52	1.31	1.38	1.03	0.90	0.69	0.55	5 300	4 800
European Spruce	8.3	6.2	5.2	4.1	3.1	6.2	4.5	3.8	3.1	2.4	1.10	0.97	0.83	1.10	0.83	0.69	0.55	0.41	5 900	3 100
Sitka Spruce	7.6	5.5	4.8	3.8	3.1	5.5	4.1	3.4	2.8	2.1	1.10	0.97	0.83	1.10	0.83	0.69	0.55	0.41	6 600	3 400
HARDWOODS																				
(a) Imported																				
Abura	13.8	10.3	9.0	6.9	5.5	10.3	7.6	6.6	5.2	4.1	2.34	2.07	1.72	2.07	1.52	1.31	1.03	0.83	8 300	4 500
African Mahogany	12.4	9.3	7.9	6.2	4.8	9.7	7.2	6.2	4.8	3.8	2.07	1.79	1.52	1.72	1.24	1.10	0.83	0.69	7 900	4 100
Afrormosia	22.1	16.5	14.1	11.0	8.6	15.9	11.7	10.3	7.9	6.2	4.14	3.59	3.10	2.62	1.93	1.65	1.31	1.03	10 300	6 900
Greenheart	37.9	28.3	24.1	19.0	15.2	27.6	20.7	17.9	13.8	11.0	6.20	5.38	4.62	4.83	3.59	3.10	2.41	1.93	17 200	12 400
Gurjun/keruing	17.2	12.8	11.0	8.6	6.9	13.8	10.3	9.0	6.9	5.5	3.10	2.76	2.34	2.34	1.72	1.52	1.17	0.90	12 400	8 300
Iroko	20.7	15.5	13.4	10.3	8.3	15.2	11.4	9.7	7.6	5.9	4.14	3.59	3.10	2.34	1.72	1.52	1.17	0.90	9 000	5 900
Jarrah	19.3	14.5	12.4	9.7	7.6	15.9	11.7	10.3	7.9	6.2	4.14	3.59	3.10	2.34	1.72	1.52	1.17	0.90	10 300	6 900
Karri	22.1	16.5	14.1	11.0	8.6	16.5	12.4	10.7	8.3	6.6	4.83	4.14	3.59	2.48	1.86	1.59	1.24	0.97	13 800	8 300
Opepe	25.5	19.3	16.5	12.8	10.3	22.1	16.5	14.1	11.0	8.6	5.52	4.83	4.14	3.10	2.28	2.00	1.52	1.24	12 400	7 600
Red Meranti/Red Seraya	12.4	9.3	7.9	6.2	4.8	9.7	7.2	6.2	4.8	3.8	1.79	1.59	1.31	1.52	1.10	0.97	0.76	0.62	7 600	4 100
Sapele	19.3	14.5	12.4	9.7	7.6	15.9	11.7	10.3	7.9	6.2	4.14	3.59	3.10	2.34	1.72	1.52	1.17	0.90	9 700	6 200
Teak	22.1	16.5	14.1	11.0	8.6	16.5	12.4	10.7	8.3	6.6	4.14	3.59	3.10	2.34	1.72	1.52	1.17	0.90	11 000	6 900
(b) Home-grown																				
European Ash	17.2	12.8	11.0	8.6	6.9	11.0	8.3	6.9	5.5	4.5	3.10	2.76	2.34	2.76	2.07	1.79	1.38	1.10	10 000	6 600
European Beech	17.2	12.8	11.0	8.6	6.9	11.0	8.3	6.9	5.5	4.5	3.10	2.76	2.34	2.76	2.07	1.79	1.38	1.10	10 000	6 600
European Oak	15.9	11.7	10.3	7.9	6.2	11.0	8.3	6.9	5.5	4.5	3.10	2.76	2.34	2.48	1.86	1.59	1.24	0.97	8 600	4 500

Note: These stresses apply to timber having a moisture content exceeding 18%.

Table 3.2 Dry stresses and moduli of elasticity

Standard name	Bending and tension parallel to the grain					Compression parallel to the grain					Compression perpendicular to the grain			Shear parallel to the grain					Modulus of elasticity for all grades	
	Basic (N/mm²)	75 Grade (N/mm²)	65 Grade (N/mm²)	50 Grade (N/mm²)	40 Grade (N/mm²)	Basic (N/mm²)	75 Grade (N/mm²)	65 Grade (N/mm²)	50 Grade (N/mm²)	40 Grade (N/mm²)	Basic (N/mm²)	75/65 Grades (N/mm²)	50/40 Grades (N/mm²)	Basic (N/mm²)	75 Grade (N/mm²)	65 Grade (N/mm²)	50 Grade (N/mm²)	40 Grade (N/mm²)	Mean (N/mm²)	Minimum (N/mm²)
SOFTWOODS																				
(a) Imported																				
Douglas Fir	18.6	13.1	11.0	8.6	6.6	14.5	10.3	8.6	6.6	5.2	2.62	2.34	1.93	1.93	1.34	1.21	0.90	0.76	11 700	6 600
Western Hemlock (unmixed)	15.9	11.4	9.3	7.6	5.9	12.4	9.3	7.9	6.2	4.8	2.07	1.72	1.52	1.65	1.21	1.07	0.83	0.66	10 000	5 900
Western Hemlock (commercial)	14.5	10.0	8.6	6.6	5.2	11.0	8.3	6.9	5.2	4.1	2.07	1.72	1.52	1.52	1.14	0.97	0.76	0.62	9 300	5 500
Parana Pine	14.5	10.0	8.6	6.6	5.2	12.4	9.3	7.9	6.2	4.8	2.21	1.93	1.65	1.65	1.21	1.07	0.83	0.60	9 000	4 800
Pitch Pine	18.6	13.1	11.0	8.6	6.6	14.5	10.3	8.6	6.6	5.2	2.62	2.34	1.93	1.93	1.34	1.21	0.90	0.76	11 700	6 600
Redwood	14.5	10.0	8.6	6.6	5.2	11.0	7.9	6.6	4.8	3.8	2.21	1.93	1.65	1.52	1.14	0.97	0.76	0.62	8 300	4 500
Whitewood	14.5	10.0	8.6	6.6	5.2	11.0	7.9	6.6	4.8	3.8	2.07	1.72	1.52	1.52	1.14	0.97	0.76	0.62	8 300	4 500
Canadian Spruce	13.8	9.7	7.9	6.2	5.2	11.0	7.9	6.6	4.8	3.8	2.07	1.72	1.52	1.52	1.14	0.97	0.76	0.62	9 000	5 500
Western Red Cedar	11.0	7.6	6.6	5.2	3.8	9.0	5.9	4.8	3.4	2.8	1.52	1.31	1.10	1.38	0.97	0.83	0.69	0.55	6 900	4 100
(b) Home-grown																				
Douglas Fir	17.9	12.4	10.7	8.3	6.6	13.8	10.0	8.3	6.2	4.8	2.48	2.21	1.93	1.52	1.14	0.97	0.76	0.62	10 000	4 800
Larch	17.2	12.1	10.3	7.9	6.2	13.1	9.3	7.6	5.5	4.5	2.62	2.34	1.93	1.72	1.21	1.07	0.83	0.66	9 700	4 800
Scots Pine	15.2	9.7	7.9	6.2	5.2	11.7	7.9	6.6	4.8	3.8	2.48	2.21	1.93	1.52	1.14	0.97	0.76	0.62	9 700	5 500
European Spruce	11.0	7.2	5.9	4.8	3.4	9.0	5.9	4.8	3.4	2.8	1.65	1.38	1.24	1.24	0.90	0.76	0.62	0.45	6 900	3 800
Sitka Spruce	10.3	6.6	5.5	4.5	3.4	8.3	5.2	4.1	3.1	2.4	1.65	1.38	1.24	1.24	0.90	0.76	0.62	0.45	7 200	3 800
HARDWOODS																				
(a) Imported																				
Abura	16.5	12.1	10.3	7.9	6.2	13.8	10.0	8.3	6.2	4.8	3.45	3.10	2.48	2.41	1.65	1.45	1.14	0.90	9 300	4 800
African Mahogany	15.2	10.7	9.0	7.2	5.5	13.1	9.3	7.6	5.5	4.5	3.10	2.62	2.21	1.93	1.34	1.21	0.90	0.76	8 600	4 500
Afrormosia	26.2	19.3	15.9	12.4	9.7	22.1	15.2	12.4	9.3	7.6	6.21	5.17	4.48	2.76	2.07	1.79	1.38	1.10	12 100	7 900
Greenheart	41.4	31.0	26.9	20.7	16.5	30.3	22.8	19.7	15.2	12.1	9.31	7.93	6.90	5.52	3.93	3.38	2.62	2.14	18 600	13 400
Gurjun/keruing	22.8	14.8	12.4	9.7	7.9	19.3	13.1	11.0	8.3	6.6	4.48	3.79	3.45	2.62	1.86	1.65	1.28	0.97	13 800	9 300
Iroko	23.4	17.6	15.2	11.7	9.3	19.3	14.5	12.1	9.0	7.2	6.21	5.17	4.48	2.62	1.86	1.65	1.28	0.97	10 300	6 900
Jarrah	23.4	16.9	14.1	11.0	8.6	20.7	15.2	12.4	9.3	7.6	6.21	5.17	4.48	2.62	1.86	1.65	1.28	0.97	12 100	7 900
Karri	26.2	19.3	15.9	12.4	9.7	22.1	15.9	13.1	9.7	7.9	7.24	6.21	5.17	2.76	2.07	1.72	1.34	1.10	15 500	9 700
Opepe	29.0	22.4	18.6	14.5	11.7	24.8	18.6	15.9	12.4	9.7	8.27	7.24	6.21	3.72	2.48	2.21	1.65	1.34	13 800	9 300
Red Meranti/Red Seraya	15.2	10.7	9.0	7.2	5.5	13.1	9.3	7.6	5.5	4.5	2.62	2.34	1.93	1.72	1.21	1.07	0.83	0.66	8 300	4 500
Sapele	23.4	16.9	14.1	11.0	8.6	20.7	15.2	12.4	9.3	7.6	6.21	5.17	4.48	2.76	1.86	1.65	1.28	0.97	11 000	6 900
Teak	26.2	19.3	15.9	12.4	9.7	22.1	15.9	13.1	9.7	7.9	6.21	5.17	4.48	2.62	1.86	1.65	1.28	0.97	12 400	7 900
(b) Home-grown																				
European Ash	22.8	14.8	12.4	9.7	7.9	15.2	10.3	8.6	6.6	5.2	4.48	3.79	3.45	3.10	2.28	2.00	1.52	1.24	11 400	7 200
European Beech	22.8	14.8	12.4	9.7	7.9	15.2	10.3	8.6	6.6	5.2	4.48	3.79	3.45	3.10	2.28	2.00	1.52	1.24	11 400	7 200
European Oak	20.7	13.8	11.7	9.0	7.2	15.2	10.3	8.6	6.6	5.2	4.48	3.79	3.45	3.10	2.07	1.72	1.34	1.10	9 700	5 200

Note: These stresses apply to timber having a moisture content exceeding 18%.

Table 3.3 Softwood species groups

Species group	*Standard name*	*Origin*
S1	Douglas Fir	Imported
	Pitch Pine	Imported
	[Douglas Fir	Home-grown
	Larch]	Home-grown
	Douglas Fir-Larch	Canada
S2	Western Hemlock (unmixed)	Imported
	Western Hemlock (commercial)	Imported
	Parana Pine	Imported
	Redwood	Imported
	Whitewood	Imported
	[Canadian Spruce	Imported
	Scots Pine]	Home-grown
	Hem-Fir	Canada
	Princess Spruce	Canada
	Western White Spruce	Canada
S3	[European Spruce	Home-grown
	Sitka Spruce]	Home-grown
	Western Red Cedar	Imported

Table 3.4 Green stresses and moduli of elasticity for grouped softwoods

Species group	*Grade*	*Bending and tension parallel to grain (N/mm²)*	*Compression parallel to grain (N/mm²)*	*Compression perpendicular to grain (N/mm²)*	*Shear parallel to grain (N/mm²)*	*Modulus of elasticity*	
						Mean (N/mm²)	*Minimum (N/mm²)*
S1	Basic	13.8	9.7	1.72	1.38		
	75	10.3	7.2	1.52	1.03		
	65	9.0	6.2	1.52	0.90	9000	4500
	50	6.9	4.8	1.31	0.69		
	40	5.5	3.8	1.31	0.55		
S2	Basic	11.0	8.3	1.38	1.38		
	75	8.3	6.2	1.17	1.03		
	65	6.9	5.2	1.17	0.90	6900	4100
	50	5.5	4.1	1.03	0.69		
	40	4.5	3.1	1.03	0.55		
S3	Basic	7.6	5.5	1.03	1.10		
	75	5.5	4.1	0.90	0.83		
	65	4.8	3.4	0.90	0.69	5900	3100
	50	3.8	2.8	0.76	0.55		
	40	3.1	2.1	0.76	0.41		

Note: These stresses apply to timber having a moisture content exceeding 18%.

Table 3.5 Dry stresses and moduli of elasticity for grouped softwoods

Species group	*Grade*	*Bending and tension parallel to grain (N/mm²)*	*Compression parallel to grain (N/mm²)*	*Compression perpendicular to grain (N/mm²)*	*Shear parallel to grain (N/mm²)*	*Modulus of elasticity*	
						Mean (N/mm²)	*Minimum (N/mm²)*
S1	Basic	17.2	13.1	2.48	1.52		
	75	12.1	9.3	2.21	1.14		
	65	10.3	7.6	2.21	0.97	9700	4800
	50	7.9	5.5	1.93	0.76		
	40	6.2	4.5	1.93	0.62		
S2	Basic	13.8	11.0	2.07	1.52		
	75	9.7	7.9	1.72	1.14		
	65	7.9	6.6	1.72	0.97	8300	4500
	50	6.2	4.8	1.52	0.76		
	40	5.2	3.8	1.52	0.62		
S3	Basic	10.3	8.3	1.52	1.24		
	75	6.6	5.2	1.31	0.90		
	65	5.5	4.1	1.31	0.76	6900	3800
	50	4.5	3.1	1.10	0.62		
	40	3.4	2.4	1.10	0.45		

Note: These stresses apply to timber having a moisture content not exceeding 18%.

The green grade stresses given in all of these tables for the four grades have been obtained by multiplying the basic stresses by approximately three-quarters, two-thirds, one-half and two-fifths, i.e. their respective strength ratios. Exceptions are modulus of elasticity and compression perpendicular to the grain or bearing strength. Modulus of elasticity is not affected to any large extent by the visual defects previously discussed and so this code allows for a strength ratio of 100% for all grades. As bearing is a local effect, influenced only by wane, the four grades are based on the wane allowance in the grading rules, which is considerably above that given by the various strength ratios.

3.2.6 BS 4978:1973

In this code, the part which deals with visual grading allots two standard grades for construction purposes – General Structural (GS) and Special Structural (SS).

The marked difference between this code and CP 112:Part 2:1971 is the method used to derive these two grade levels. The method is known as the 'knot area ratio' (KAR) and differs from the previous code in that it recognizes the large part that is played by the knot in reducing strength in regard to its three-dimensional (volumetric) shape rather than its surface area (cut) shape.

Table 3.6 Green stresses and moduli of elasticity

Standard name	Bending		Tension		*Compression parallel to grain*		*Compression perpendicular to grain*		*Shear parallel to grain*		*Modulus of elasticity*			
											Mean		*Minimun*	
	SS (N/mm^2)	*GS* (N/mm^2)	*SS* (N/mm^2)	*GS* (N/mm^2)	*SS* (N/mm^2)	*GS* (N/mm^2)	*SS* (N/mm^2)	*GS* (N/mm^2)	*SS* (N/mm^2)	*GS* (N/mm^2)	*SS* (N/mm^2)	*GS* (N/mm^2)	*SS* (N/mm^2)	*GS* (N/mm^2)
(a) Imported														
Douglas Fir-Larch	7.6	5.3	5.3	3.7	7.2	5.0	1.35	1.20	0.86	0.86	11 300	10 200	6 600	6 000
Hem-Fir	5.9	4.0	4.0	2.9	5.9	4.0	1.03	0.92	0.70	0.70	9 500	8 500	5 900	5 200
Parana Pine	5.9	4.0	4.0	2.9	6.7	4.6	1.14	1.01	0.76	0.76	9 500	8 500	5 400	4 800
Pitch Pine	7.6	5.3	5.3	3.7	7.2	5.0	1.35	1.20	0.86	0.86	11 900	10 700	7 800	7 000
Redwood	5.9	4.0	4.0	2.9	5.4	3.7	1.14	1.01	0.70	0.70	8 200	7 000	4 700	4 000
Whitewood	5.9	4.0	4.0	2.9	5.4	3.7	1.03	0.92	0.70	0.70	8 200	7 000	4 700	4 000
Spruce-Pine-Fir	5.5	3.8	3.8	2.7	5.4	3.7	1.03	0.92	0.70	0.70	8 000	7 200	4 300	3 900
Western Red Cedar	4.5	3.2	3.2	2.2	4.0	2.8	0.77	0.69	0.62	0.62	7 400	6 700	5 000	4 600
(b) Home-grown														
Douglas Fir	7.3	5.1	5.1	3.6	6.7	4.6	1.29	1.15	0.70	0.70	9 800	8 800	4 700	4 200
Larch	6.9	4.8	4.8	3.4	6.3	4.4	1.35	1.20	0.86	0.86	9 200	8 200	4 400	4 000
Scots Pine	5.5	3.8	3.8	2.7	5.4	3.7	1.29	1.15	0.70	0.70	9 000	8 100	4 300	3 900
European Spruce	4.2	2.9	2.9	2.0	4.0	2.8	0.82	0.73	0.55	0.55	6 100	5 500	3 200	2 900
Sitka Spruce	3.8	2.7	2.7	1.9	3.6	2.5	0.82	0.73	0.55	0.55	6 500	5 900	3 100	2 800

Note: These stresses apply to timber having a moisture content exceeding 18%. These stresses also apply respectively to Douglas Fir, Western Hemlock (commercial) and Canadian Spruce as designated in Table 1.1 when graded to the visual grades in BS 4978. Where appropriate (see Clause 3.6.1, CP 112) the basic stresses in compression perpendicular to grain, given in Tables 3.4 and 3.5 for these species may be used for the above species groups.

Table 3.7 Dry stresses and moduli of elasticity

Standard name	*Bending*		*Tension*		*Compression parallel to grain*		*Compression perpendicular to grain*		*Shear parallel to grain*		*Modulus of elasticity*			
											Mean		*Minimun*	
	SS (N/mm^2)	*GS* (N/mm^2)	*SS* (N/mm^2)	*GS* (N/mm^2)	*SS* (N/mm^2)	*GS* (N/mm^2)	*SS* (N/mm^2)	*GS* (N/mm^2)	*SS* (N/mm^2)	*GS* (N/mm^2)	*SS* (N/mm^2)	*GS* (N/mm^2)	*SS* (N/mm^2)	*GS* (N/mm^2)
(a) Imported														
Douglas Fir-Larch	9.3	6.5	6.5	4.6	9.6	6.7	2.00	1.80	0.96	0.96	12 500	11 300	7 300	6 600
Hem-Fir	7.3	5.1	5.1	3.5	7.9	5.5	1.55	1.38	0.80	0.80	10 700	9 600	6 600	5 900
Parana Pine	7.3	5.1	5.1	3.5	9.1	6.3	1.71	1.52	0.90	0.90	10 500	9 400	6 000	5 300
Pitch Pine	9.3	6.5	6.5	4.6	9.1	6.3	2.00	1.80	0.94	0.94	12 900	11 600	8 500	7 600
Redwood	7.3	5.1	5.1	3.5	8.0	5.6	1.71	1.52	0.86	0.86	10 000	8 600	5 700	4 900
Whitewood	7.3	5.1	5.1	3.5	8.0	5.6	1.55	1.38	0.86	0.86	10 000	8 600	5 700	4 900
Spruce-Pine-Fir	6.9	4.8	4.8	3.3	7.0	4.8	1.55	1.38	0.79	0.79	8 900	8 000	4 800	4 300
Western Red Cedar	5.2	3.6	3.6	2.5	5.9	4.1	1.16	1.03	0.70	0.70	8 000	7 200	5 500	4 900
(b) Home-grown														
Douglas Fir	9.0	6.3	6.3	4.4	9.0	6.3	1.93	1.72	0.87	0.87	10 700	9 800	5 200	4 700
Larch	8.6	6.0	6.0	4.2	8.6	6.0	2.00	1.80	1.00	1.00	10 600	9 500	5 100	4 600
Scots Pine	7.6	5.3	5.3	3.7	8.0	5.6	1.93	1.72	0.86	0.86	10 600	9 500	5 100	4 600
European Spruce	5.5	3.9	3.9	2.7	5.6	3.9	1.24	1.10	0.71	0.71	7 500	6 800	4 000	3 600
Sitka Spruce	5.2	3.6	3.6	2.5	5.0	3.5	1.24	1.10	0.66	0.66	7 800	7 000	3 700	3 300

Note: These stresses apply to timber having a moisture content not exceeding 18%. These stresses also apply respectively to Douglas Fir, Western Hemlock (commercial) and Canadian Spruce as designated in Table 1.1 when graded to the visual grades in BS 4978. Where appropriate (see Clause 3.6.1, CP 112) the basic stresses in compression perpendicular to grain, given in Tables 3.4 and 3.5 for these species may be used for the above species groups.

Table 3.8 Green stresses and moduli of elasticity for grouped softwoods

Species group	Grade	Bending (N/mm²)	Tension (N/mm²)	Compression parallel to grain (N/mm²)	Compression perpendicular to grain (N/mm²)	Shear parallel to grain (N/mm²)	Modulus of elasticity Mean (N/mm²)	Modulus of elasticity Minimum (N/mm²)
S1	SS	6.9	4.8	6.3	1.29	0.70	9 200	4 400
	GS	4.8	3.4	4.4	1.15	0.70	8 200	4 000
S2	SS	5.5	3.8	5.4	1.03	0.70	8 000	4 300
	GS	3.8	2.7	3.7	0.92	0.70	7 000	3 900
S3	SS	3.8	2.7	3.6	0.77	0.55	6 100	3 100
	GS	2.7	1.9	2.5	0.69	0.55	5 500	2 800

Note: These stresses apply to timber having a moisture content exceeding 18%.

Another important improvement is the provision for all timbers to be marked as being of either GS or SS quality. This demands that the person providing the stamp marking should be suitably trained and examined prior to receiving this important appointment. Such training schemes as exist today are provided under the auspices of the British Standards Institution.

The design stresses for these two grades and various species are reproduced in Tables 3.6 and 3.7 for green and dry conditions, respectively. They are taken from Addendum No. 1, 1973 to CP 112:Part 2:1971.

As in the case of the numerical grades, grouping of timber is also used in GS and SS grades and Tables 3.8 and 3.9 should be referred to for the stress levels.

Table 3.9 Dry stresses and moduli of elasticity for grouped softwoods

Species group	Grade	Bending (N/mm²)	Tension (N/mm²)	Compression parallel to grain (N/mm²)	Compression perpendicular to grain (N/mm²)	Shear parallel to grain (N/mm²)	Modulus of elasticity Mean (N/mm²)	Modulus of elasticity Minimum (N/mm²)
S1	SS	8.6	6.0	8.6	1.93	0.87	10 600	5 100
	GS	6.0	4.2	6.0	1.72	0.87	9 500	4 600
S2	SS	6.9	4.8	7.0	1.55	0.79	8 900	4 800
	GS	4.8	3.3	4.8	1.38	0.79	8 000	4 300
S3	SS	5.2	3.6	5.0	1.16	0.66	7 500	3 700
	GS	3.6	2.5	3.5	1.03	0.66	6 800	3 300

Note: These stresses apply to timber having a moisture content not exceeding 18%.

3.3 Mechanical stress grading

Commonly referred to as machine stress grading this is a relatively new introduction into this country but it has brought with it a much more sophisticated and accurate approach to grading than can be said of the two visual methods.

Visual grading has the disadvantage of not being able to see the main body content of the timber throughout the length of the piece and therefore it cannot separate the naturally weak from the naturally strong. It is therefore possible for a piece of timber with many apparent defects to be stronger than a similar piece with no defects. On the other hand, machine grading makes allowance for the combined effects of visual defects and hidden density giving a more accurate strength grade to any given piece leading to more economical usage.

Mechanical stress grading examines the relationship which exists between the deflection of a piece of timber under a small concentrated load and its breaking strength. It has been shown by tests that timber can be machine graded on the flat to estimate its strength in the depth and it is around this method that current machines have been designed. Basically the piece of timber to be graded passes across three rollers, two static and one moving. The static rollers are at a known distance apart and the moving roller applies a small concentrated central load. The machine must be approved and calibrated by the British Standards Institution and periodic unannounced visits are made by inspectors to check those calibrations.

BS 4978:1973 allows for the machine stress grading of GS and SS and they carry the prefix M to give MGS and MSS as the recognized approved marking. In addition, the British Standard kite mark and approval number is stamped on each piece of timber graded. Reference may be made to Tables 3.10 and 3.11 for the respective stresses allotted to these two particular machine grades for a limited amount of species.

There are also two other machine stress grade values published in Addendum No. 1 and these are designated M50 and M75. Again they are monitored by the British Standards Institution and carry the appropriate markings but they are also easily recognized by the blue and red periodic dashes of dye marks applied to the overall length of each piece as it passes through the machine. Blue defines M50 and red M75. Reference should be made to Tables 3.10 and 3.11 (as for MGS and MSS) for the derived stress values. Therefore, summarizing the grade stresses currently available to the designers we have:

Visual – 40, 50, 65, 75, GS and SS
Machine – MGS, MSS, M50 and M75

It should be noted that stress grade values for hardwoods are as yet only available in the numerical visual grades.

Table 3.10 Green stresses and moduli of elasticity for machine stress-graded softwoods

Property	*Grade*	*Standard name*					
		Western Hemlock (commercial) (N/mm²)	*Redwood and Whitewood (N/mm²)*	*Douglas Fir (home-grown) (N/mm²)*	*Scots Pine (home-grown) (N/mm²)*	*Sitka Spruce (home-grown) (N/mm²)*	*Western White Spruce (N/mm²)*
Bending	M75	8.1	8.1	10.0	7.0	4.9	7.6
	M50	5.3	5.3	6.7	4.5	3.3	5.0
	MSS	5.9	5.9	7.3	5.5	3.8	5.5
	MGS	4.0	4.0	5.1	3.8	2.7	3.8
Tension	M75	5.7	5.7	7.0	4.8	3.4	5.3
	M50	3.7	3.7	4.7	3.1	2.3	3.5
	MSS	4.0	4.0	5.1	3.8	2.7	3.8
	MGS	2.9	2.9	3.6	2.7	1.9	2.7
Compression parallel to grain	M75	7.9	7.3	9.2	6.7	4.6	7.3
	M50	5.2	4.8	6.1	4.3	3.1	4.8
	MSS	5.9	5.4	6.7	5.4	3.6	5.4
	MGS	4.0	3.7	4.6	3.7	2.5	3.7
Compression perpendicular to grain	M75	1.20	1.20*	1.50	1.50	0.97	1.20
	M50	1.03	1.03*	1.29	1.29	0.82	1.03
	MSS	1.03	1.03*	1.29	1.29	0.82	1.03
	MGS	0.92	0.92*	1.15	1.15	0.73	0.92
Shear parallel to grain	M75	1.03	1.03	1.03	1.03	0.82	1.03
	M50	0.70	0.70	0.70	0.70	0.55	0.70
	MSS	0.70	0.70	0.70	0.70	0.55	0.70
	MGS	0.70	0.70	0.70	0.70	0.55	0.70
Mean modulus of elasticity	M75	10 600	8 700	11 300	10 300	7 500	9 200
	M50	9 300	7 300	9 500	9 000	6 400	7 700
	MSS	9 600	8 300	9 700	9 600	6 800	8 300
	MGS	8 700	7 200	8 600	8 700	6 000	7 400
Minimum modulus of elasticity	M75	7 100	5 500	7 100	6 800	4 600	5 800
	M50	6 200	4 500	6 200	6 000	4 000	4 900
	MSS	6 400	5 200	6 400	6 400	4 200	5 200
	MGS	5 800	4 400	5 800	5 800	3 700	4 600

Note: These stresses apply to timber having a moisture content exceeding 18%.
* Where redwood is used separately these stresses may be increased by a factor of 1.1.

Table 3.11 Dry stresses and moduli of elasticity for machine stress-graded softwoods

Property	*Grade*	*Standard name*					
		Western Hemlock (commercial) (N/mm²)	*Redwood and Whitewood (N/mm²)*	*Douglas Fir (home-grown) (N/mm²)*	*Scots Pine (home-grown) (N/mm²)*	*Sitka Spruce (home-grown) (N/mm²)*	*Western White Spruce (N/mm²)*
Bending	M75	10.0	10.0	12.4	9.6	6.6	9.7
	M50	6.6	6.6	8.3	6.2	4.5	6.2
	MSS	7.3	7.3	9.0	7.6	5.2	6.9
	MGS	5.1	5.1	6.3	5.3	3.6	4.8
Tension	M75	7.0	7.0	8.7	6.7	4.6	6.7
	M50	4.6	4.6	5.8	4.3	3.2	4.3
	MSS	5.1	5.1	6.3	5.3	3.6	4.8
	MGS	3.5	3.5	4.4	3.7	2.5	3.3
Compression parallel to grain	M75	10.8	10.8	12.4	10.0	6.4	9.5
	M50	7.1	7.1	8.3	6.5	4.3	6.3
	MSS	7.9	8.0	9.0	8.0	5.0	7.0
	MGS	5.5	5.6	6.3	5.6	3.5	4.8
Compression perpendicular to grain	M75	1.80	1.80*	2.25	2.25	1.45	1.80
	M50	1.55	1.55*	1.93	1.93	1.24	1.55
	MSS	1.55	1.55*	1.93	1.93	1.24	1.55
	MGS	1.38	1.38*	1.72	1.72	1.10	1.38
Shear parallel to grain	M75	1.19	1.28	1.30	1.28	0.98	1.19
	M50	0.80	0.86	0.87	0.86	0.66	0.79
	MSS	0.80	0.86	0.87	0.86	0.66	0.79
	MGS	0.80	0.86	0.87	0.86	0.66	0.79
Mean modulus of elasticity	M75	12 000	10 700	12 500	12 000	9 000	10 200
	M50	10 500	9 000	10 500	10 500	7 700	8 600
	MSS	10 800	10 200	10 800	11 200	8 200	9 200
	MGS	9 800	8 800	9 500	10 200	7 200	8 200
Minimum modulus of elasticity	M75	8 000	6 700	8 000	8 000	5 500	6 400
	M50	7 000	5 500	7 000	7 000	4 800	5 400
	MSS	7 200	6 400	7 200	7 500	5 100	5 800
	MGS	6 500	5 400	6 500	6 800	4 500	5 100

Note: These stresses apply to timber having a moisture content not exceeding 18%.

3.4 Timber stress graded in Canada

It will be recalled that in the chapter dealing with materials, we covered those timbers imported from Canada. Naturally, they too carry a stress grade but unlike the stress grades previously discussed, they are derived in the country of origin and conform to Canadian standards.

The particular standard in question is the 'Structural joists and planks section of the National Lumber Grades Authority 1970 standard grading rules

Table 3.12 Grade stresses and moduli of elasticity of Canadian Douglas Fir-Larch

Type of stress or modulus	*Values of stress or modulus*							
	Green				*Dry*			
	Select structural (N/mm²)	*No. 1 (N/mm²)*	*No. 2 (N/mm²)*	*No. 3 (N/mm²)*	*Select structural (N/mm²)*	*No. 1 (N/mm²)*	*No. 2 (N/mm²)*	*No. 3 (N/mm²)*
Bending	10.0	8.5	6.8	4.0	12.3	10.4	8.4	4.8
Tension	7.0	6.0	4.8	2.8	8.6	7.3	5.9	3.4
Compression parallel to grain	7.6	6.8	5.7	3.6	10.2	9.2	7.7	4.9
Compression perpendicular to grain	1.35	1.35	1.20	0.90	2.00	2.00	1.80	1.35
Shear parallel to grain	0.86	0.86	0.86	0.77	0.96	0.96	0.96	0.86
Mean modulus of elasticity	12 500	11 700	10 900	9 700	13 800	13 000	12 100	10 700
Minimum modulus of elasticity	7 300	6 900	6 400	5 700	8 100	7 600	7 100	6 300

Applicable to sections of nominal 50 mm to 100 mm thickness and nominal 150 mm, or greater, width.

The stresses for the green condition apply to timber having a moisture content exceeding 18% and for the dry condition to timber having a moisture content not exceeding 18%.

Where members are free from wane the basic stress in compression perpendicular to grain may be used in design, irrespective of their actual grade. The basic stress values are 1.79 N/mm² for the green condition and 2.62 N/mm² for the dry condition.

for Canadian lumber'. In specification documents this is frequently referred to as 'graded to NGLA standards'. CP 112:Part 2:1971, Clause 2.3.2 allows for the use of this timber without regrading.

The stress grades for Canadian timber are given in Appendix B of the code and are classified for Douglas Fir-Larch, Hem-Fir and Spruce-Pine-Fir and the assigned values are reproduced in Tables 3.12, 3.13 and 3.14. These tables indicate four different classifications for the stress values in accordance with differing qualities and they are Select Structural, No. 1, No. 2 and No. 3. However, it has previously been indicated that currently the predominant usage of Canadian timber is S-P-F in timber frame housing construction and in

Table 3.13 Grade stresses and moduli of elasticity of Canadian Hem-Fir

Type of stress or modulus	*Values of stress or modulus*							
	Green				*Dry*			
	Select structural (N/mm²)	*No. 1 (N/mm²)*	*No. 2 (N/mm²)*	*No. 3 (N/mm²)*	*Select structural (N/mm²)*	*No. 1 (N/mm²)*	*No. 2 (N/mm²)*	*No. 3 (N/mm²)*
Bending	7.7	6.6	5.3	3.0	9.6	8.1	6.5	3.8
Tension	5.4	4.6	3.7	2.1	6.7	5.7	4.6	2.6
Compression parallel to grain	6.2	5.6	4.7	3.0	8.4	7.5	6.4	4.0
Compression perpendicular to grain	1.03	1.03	0.92	0.69	1.55	1.55	1.38	1.03
Shear parallel to grain	0.70	0.70	0.70	0.61	0.80	0.80	0.80	0.70
Mean modulus of elasticity	10 400	9 800	9 100	8 100	11 800	11 100	10 300	9 100
Minimum modulus of elasticity	6 400	6 000	5 600	5 000	7 300	6 800	6 400	5 600

Applicable to sections of nominal 50 mm to 100 mm thickness and nominal 150 mm, or greater, width.

The stresses for the green condition apply to timber having a moisture content exceeding 18% and for the dry condition to timber having a moisture content not exceeding 18%.

Where members are free from wane the basic stress in compression perpendicular to grain may be used in design, irrespective of their actual grade. The basic stress values are 1.38 N/mm² for the green condition and 2.07 N/mm² for the dry condition.

this respect the classifications of grades are somewhat different for this species, from those given in the code.

Information issued by the Council of Forest Industries of British Columbia (COFI) gives five stress grades values, namely, No. 1, No. 2, Construction, Standard and Stud for kiln-dried moisture conditions. These are reproduced in Table 3.15 and comparison with the dry stresses in Table 3.14 shows that Select Structural is not covered, No. 1 and No. 2 are the same, Construction and Standard are not covered and the Stud values approximate closely to No. 3. When designing for the use of CLS(S-P-F) in timber frame housing the grades most frequently used are No. 1, No. 2 and Construction.

3.5 Stress grades for laminated timber

A distinct advantage of laminated timber over basic solid sections of similar species is that it enables structural members of virtually any length and cross-section to be produced from relatively small pieces of timber.

With laminated construction the timber must first be dried to a moisture content of 15% before it can be glue bonded. This is best carried out on thin

Table 3.14 Grade stresses and moduli of elasticity of Canadian Spruce-Pine-Fir

Type of stress or modulus	*Values of stress or modulus*							
	Green				*Dry*			
	Select structural (N/mm²)	*No. 1 (N/mm²)*	*No. 2 (N/mm²)*	*No. 3 (N/mm²)*	*Select structural (N/mm²)*	*No. 1 (N/mm²)*	*No. 2 (N/mm²)*	*No. 3 (N/mm²)*
Bending	7.3	6.2	5.0	2.9	9.1	7.7	6.2	3.6
Tension	5.1	4.3	3.5	2.0	6.4	5.4	4.3	2.5
Compression parallel to grain	5.7	5.1	4.3	2.7	7.4	6.7	5.6	3.6
Compression perpendicular to grain	1.03	1.03	0.91	0.69	1.55	1.55	1.38	1.03
Shear parallel to grain	0.70	0.70	0.70	0.55	0.79	0.79	0.79	0.63
Mean modulus of elasticity	8 800	8 200	7 700	6 800	9 800	9 300	8 600	7 600
Minimum modulus of elasticity	4 800	4 400	4 200	3 700	5 300	5 000	4 700	4 100

Applicable to sections of nominal 50 mm to 100 mm thickness and nominal 150 mm, or greater, width.

The stresses for the green condition apply to timber having a moisture content exceeding 18% and for the dry condition to timber having a moisture content not exceeding 18%.

Where members are free from wane the basic stress in compression perpendicular to grain may be used in design, irrespective of their actual grade. The basic stress values are 1.38 N/mm² for the green condition and 2.07 N/mm² for the dry condition.

sections to prevent severe internal stresses developing under normal fluctuations in moisture content and so a thickness of 44 mm should not be exceeded. The properties of wood are not altered by laminating and the strength is not affected by the thickness of the laminations used in construction. Basic stresses for laminated timber can therefore be derived in the same manner as previously described for green solid sections with adjustment to the dry (18%) condition.

There is a marked difference in establishing grade stresses for laminated timbers in that the grades are first specified and the strength ratios determined subsequently. Thus basic stresses are first determined and then rules for stress grading are introduced and the strength ratios for these grades and combinations of them are derived.

By far the most important influence on the grade strength of laminated timber is the effect of the size and position of knots. The strength ratio, appropriate to a grade which will depend upon the influence of the largest knot concentration likely to be encountered, must therefore be related to the number of laminations in the member.

The stress grades are designated in three separate categories, namely, LA,

Table 3.15
(a) Grade stresses for CLS Spruce-Pine-Fir (N/mm²)*

	NLGA Category/Grade				
	Structural light framing; structural joists and planks†		*Light framing*‡		*Studs*
Unit stress	*No. 1*	*No. 2*	*Const.*	*Std.*	*Stud*
Bending	7.7	6.2	5.4	3.0	4.1
Tension	5.4	4.3	3.1	1.8	2.5
Compression parallel to grain	6.7	5.6	6.0	5.0	3.2
Compression perpendicular to grain§	1.55	1.38	1.55	1.38	1.03
Shear parallel to grain	0.79	0.79	0.79	0.79	0.63
Mean modulus of elasticity	9 300	8 600	7 900	7 100	7 600
Minimum modulus of elasticity	5 000	4 700	4 300	3 900	4 100

Notes:
* Stresses are for dry timber, i.e. with a moisure content of 18% or less, and have been derived from British Standard Code of Practice CP112, Part 2:1971, Amendment Slip No. 1; and BRE Information Sheet No. IS 5/75.
† Applicable to timber with a nominal thickness of 38 mm and a width of 89 mm or greater. For smaller sections, the factors indicated below must be applied.
‡ Applicable to timber in the Light Framing category with a thickness of 38 mm and a width of 89 mm. For smaller sections, the factors given below must be applied.
§ Where the members are free from wane, the basic dry stress value of 2.07 N/mm² may be used irrespective of the grade of the timber.

(b) Grade stress modification factors for nominal 2 × 2 and 2 × 3 timber

NLGA category	*Grade*	*Type of stress*	*Factor*	
			38 × 38	38 × 64
Structural Light Framing	No. 1	Bending	1.10	1.10
		Compression parallel to grain	1.19	1.19
	No. 2	Bending	1.10	1.10
		Compression parallel to grain	1.35	1.35
		Tension parallel to grain	1.00	1.06
		Bending	.90	.82

LB and LC, and these are related directly to the size of knot occurring in relation to a specified depth of total lamination. LA is the top grade available.

Slope of grain is another important factor and restrictions are made appropriate to one of the three grades to be specified.

Other natural defects are taken into account in the grading rules and are similar to those defects previously described for solid sawn section timbers.

The design stresses are finally determined by taking the basic stresses for the species chosen (see Tables 3.1 and 3.2) and multiplying by various modification factors given in CP 112:Part 2:1971. These reduction factors are related to the grade chosen and the number of laminates used.

Another strength-reducing factor in the production of laminated sections is the introduction of end joints within the laminates to obtain the finished overall length of the member. Efficiency ratings are specified for scarfed joints based upon their slope and similarly where finger jointing is used efficiency and frequency are taken into account.

Finally, the grading rules which may be used for laminated construction are covered in either Appendix A of CP 112:Part 2:1971 or BS 4978:1973.

4

Durability and treatment

4.1 Fungal and insect attack

Nearly all timbers are subject to fungal decay and insect attack to some degree and in the United Kingdom, where the climate is mild, certain insects and fungi can and do attack timber.

Blue-stain or sap-stain is the result of a staining fungus which discolours the wood, particularly pine sapwood; however, it may be safely used for structural work, provided that all other features are acceptable.

Fungi grow only on damp wood and, generally, the moisture content has to be in excess of 20%; therefore, if after conversion from the log the sections are properly stacked and seasoned, fungal attack can be monitored and controlled. The preservation of timber against fungal and insect attack is left very much to the discretion of the specifier except in so far as the building regulations Part B stipulate that there must be protection against the marine boring insect.

Timber may be protected against biological attack by either good design detail or the specification of a particularly durable species; however such methods should not be wholly relied upon where there is a definite risk involved. In this country, where more softwood is used than hardwood, the need for treatment should be constantly considered by the designer with all projects undertaken. The sapwood of all species is taken to be non-durable whilst the heartwood varies considerably in the extent of its durability. Woodboring insects are also generally confined in activity to the sapwood. Only a small number of species are resistant to marine boring attack.

4.2 Treatment

There are a number of different methods of treatment available but the most common used for structural timbers are the double vacuum treatment which makes use of organic solvents and the pressure impregnation treatment where copper/chrome/arsenic salts are used. Table 4.1 lists the various treatments available and refers to the relevant specifications.

Table 4.1 Preservative treatments and relevant specifications

Treatment	*Composition*	*Application*
Pressure treatment		
creosote	BS 144	BS 913
copper/chrome/arsenic (CCA) salts	BS 4072	BS 4072
Boron diffusion treatment*	BWPA 105	BWPA 105
Double vacuum treatment		
organic solvent (OS) preservatives	BWPA 112 to 116	BWPA 112 to 116
pentachlorophenol (PCP) in heavy oil	BWPA 117	BWPA 117
Immersion treatment		
creosote	BS 3051	
organic solvent (OS) preservatives (without contact insecticide)	BS 5056 BWPA 104 BWPA 109 BWPA 111	
organic solvent (OS) preservatives (with contact insecticide)	BWPA 108 BWPA 110 BWPA 118	

* Freshly felled timber is treated at source with *di*sodium octaborate. Seasoned timber cannot be treated by this process.

Recommendations for the preservative treatment of structural timber given in BS 5268:Part 5:1977 and the durability of those species of timber listed in CP 112:Part 2, together with their degree of susceptibility to treatment are given in Table 4.2. Finally, it is the opinion of the author that, for the relatively small additional cost involved, preservative treatment should always be recommended unless there is some other factor which makes its use unnecessary.

Table 4.2 Durability and treatability characteristics of solid timbers named in CP 112:Part 2

Common name	*Botanical name*	*Durability* (heartwood only)*	*Treatability†*	
			Heart	*Sap*
Douglas Fir (imported)	*Pseudotsuga menziesii*	MD	R	NK
Douglas Fir (home grown)	*Pseudotsuga menziesii*	ND	R	MR
‡Douglas Fir-Larch	*Pseudotsuga menziesii* *Larix occidentalis*	MD	R	NK
‡Hem-Fir	*Tsuga heterophylla Abies amabilis* *Tsuga mertensiana Abies grandis*	ND	R	
Larch	*Larix decidua* *Larix leptolepis*	MD	R	MR
Parana Pine	*Araucaria angustifolia*	ND	MR	P

Table 4.2 (continued)

Common name	*Botanical name*		*Durability* (heartwood only)*	*Treatability†*	
				Heart	*Sap*
Pitch Pine	*Pinus palustris*		D	R	P
	Pinus elliottii				
	Pinus caribaea				
Redwood or Scots Pine	*Pinus sylvestris*		ND	MR	P
‡Spruce-Pine-Fir	*Picea glauca*	*Pinus contorta*	ND	R	
	Picea engelmannii	*Pinus banksiana*			
	Picea rubens	*Abies balsamea*			
	Picea mariana	*Abies lasiocarpa*			
Western Red Cedar	*Thuja plicata*		D	R	
Whitewood or European Spruce (home grown)	*Picea abies*		ND	R	
Sitka Spruce (home grown)	*Picea sitchensis*		ND	R	MR
Whitewood	*Picea abies*		ND	R	
Greenheart	*Ocotea rodiaei*		VD	ER	NK
Gurjun/keruing	*Dipterocarpus* spp.		MD	MR-R	MR
Iroko	*Chlorophora excelsa*		VD	ER	P
Jarrah	*Eucalyptus marginata*		VD	ER	P
Karri	*Eucalyptus diversicolor*		D	ER	P
European Oak	*Quercus robur*		D	ER	P
Opepe	*Nauclea diderrichii*		VD	MR	P

* Durability classes (for explanation see PRL Technical Note No. 40): VD very durable, D durable, MD moderately durable, ND non-durable.

† Treatability classes (for explanation see PRL Bulletin No. 54): ER extremely resistant, R resistant, MR moderately resistant, P permeable, NK not known.

‡ Timbers imported under these species group names have been characterized by the least durable and most resistant species of the group.

Note: A separate durability classification for plywood has not been established. The so-called exterior grade refers primarily to the suitability of the glue bond for exterior situations. In general, durability ratings (i.e. resistance to decay) given for solid timber can be taken as applying to plywood. Plywood and solid timber sometimes differ in treatability due to the physical changes caused by the peeling process and board manufacture. Treatability ratings for solid timber cannot be simply extended to plywood.

5

The Building Regulations in timber design

5.1 General

The introduction of the Building Regulations in 1965 heralded a new and irreversible change to the administration and the control of building design and construction in this country. It was in many respects an historic event destined to change the face of the building industry.

The Building Regulations and the Building Standards (Scotland) Consolidation Regulations are to be found in all offices where there lies an interest or involvement in the art of building. Indeed no one person, company or organization can proceed in the execution of the design and construction of a building without having to comply in some respect with these regulations. They are mandatory in that they are laid as matters before Parliament and passed under the Law of the Public Health Act for England, Wales and Northern Ireland and the Building (Scotland) Act for Scotland. Under no circumstances can they be ignored. The administration and interpretation of these regulations are vested in the various local authorities throughout the land and, in the event of serious dispute between the local authority and those submitting for permission to build, there are set procedures for appeal.

The timber designer, like all others who make use of an acquired skill, particularly where the safety of the public is involved, owes a duty of care in law when exercising this skill. Consequently, it is essential that all building designers, no matter their speciality, should have at least a basic understanding of these regulations.

In regard to the media of timber design, we will discuss in general terms those subjects contained in the Building Regulations 1976 where they directly affect the structural designer. Those regulations for Scotland are very similar but particular reference should be made to the current Building Standards for detailed comparison.

5.2 Part A – Interpretation and general

5.2.1 Regulation A1

Regulation A1 enforces the title and date of commencement of the whole of the Building Regulations and the designer must comply with the date.

Regulations A2 and A3 relate to any transitional periods and revocation of previously issued regulations, respectively.

Regulation A4(g) allows for flexibility in that even though reference is made to a deemed-to-satisfy provision, it does not necessarily follow that it will be enforced if other means are shown to be satisfactory.

5.2.2 Regulation A12

Regulation A12 recognizes that there will be occasions when, for various reasons, particular sections cannot be complied with and so this regulation allows the applicant to seek a dispensation or relaxation. This dispensation or relaxation is administered by the local authority in accordance with regulation A13.

5.2.3 Regulation A15

Regulation A15 permits the local authority to examine any material which they consider to be suspect in regard to its fitness for the purpose proposed. The designer, when proposing the use of a little known material, would be well advised to seek independent certification such as is given by the British Board of Agrément.

5.3 Part B – Materials

5.3.1 Regulation B1

Regulation B1 covers the fitness of materials in regard to their acceptance within the body of the regulations whilst B2 gives the deemed-to-satisfy provision to B1. In this respect, CP 112:Part 2:1971 'The structural use of timber' would satisfy this part of the regulations. In particular, Section 2 entitled 'Materials, appliances and components'.

5.3.2 Regulation B3

Regulation B3 defines those districts where it is considered that steps should be taken to treat softwood against the possible infestation of the longhorn beetle (*Hyletrupes bajulus*).

5.3.3 Regulation B4

Regulation B4 gives the deemed-to-satisfy provision for B3 and all of the better-known proprietary systems are able to provide suitable treatment in the form of double vacuum pressure impregnation (reference should be made to BS 5268:Part 5:1977 'Preservative treatments for constructional timber').

5.4 Part C – Preparation of site and resistance to moisture

5.4.1 Regulation C3

Regulation C3 defines the need for suitable protection of floors next to the ground to prevent the passage of moisture and vapour and C4 gives the deemed-to-satisfy provisions which apply to suspended timber floors. All details should comply with these minimum standards and there is some descriptive literature on this subject, notably the Timber Research and Development Association (TRADA) have published useful details.

5.4.2 Regulations C5, C6, C7, C8 and C9

These regulations are best expanded upon in the context of timber usage by the illustrations given in Figs 5.1, 5.2, 5.3, 5.4 and 5.5.

5.5 Part D – Structural stability

5.5.1 Regulation D1

Regulation D1 deals with the meanings attributed to dead load, imposed load and wind load and it is considered that these are selfexplanatory requiring no further clarification.

5.5.2 Regulation D2

Regulation D2 gives those codes of practice which have to be used to determine the loads described in D1 and these are:

(1) as described in the regulations
(2) (a) dead loads to CP 3, Chapter V, Part 1:1967
    (b) imposed loads to CP 3, Chapter V, Part 1:1967

Although the regulations allow for other consideration of loadings, we will constantly refer to CP 3, Chapter V, Part 1:1967 because the vast majority of cases fall within the parameters of this code.

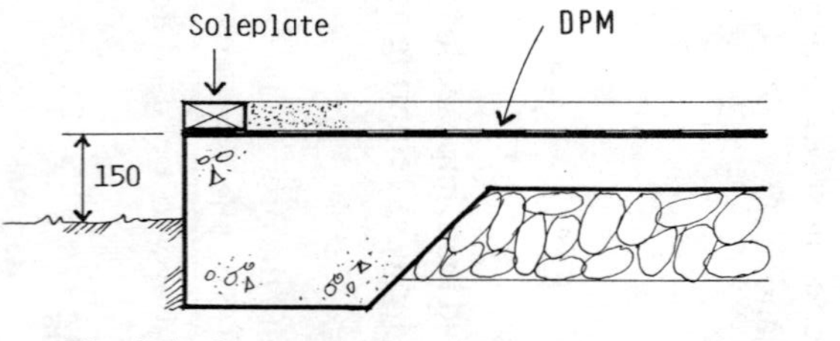

Fig. 5.1 Interpretation of Regulation C5

Fig. 5.3 Interpretation of Regulation C8

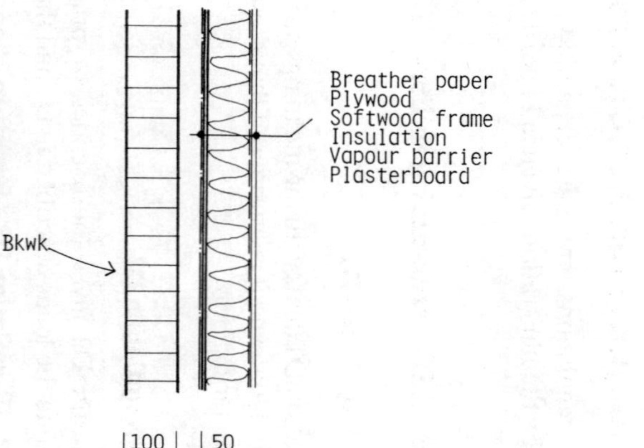

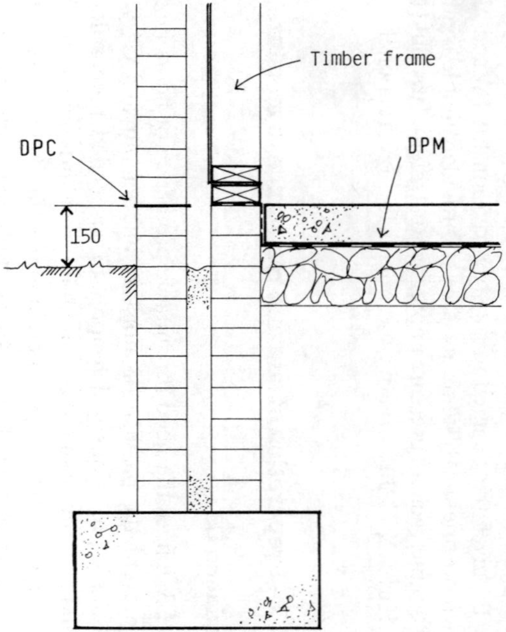

Fig. 5.2 Interpretation of Regulations C6, C7, C9(1)

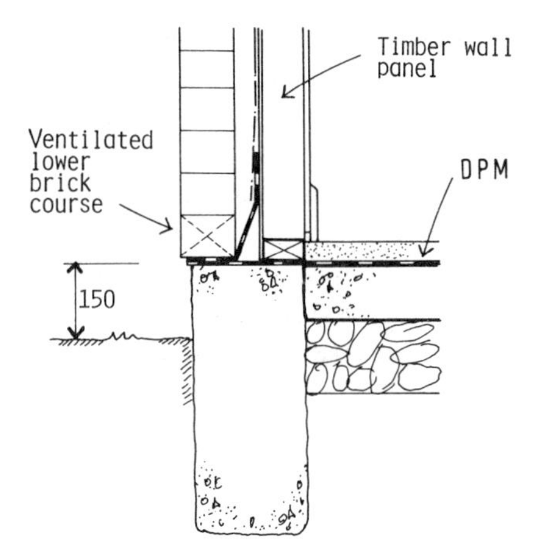

Fig. 5.4 Interpretation of Regulation C9(2)

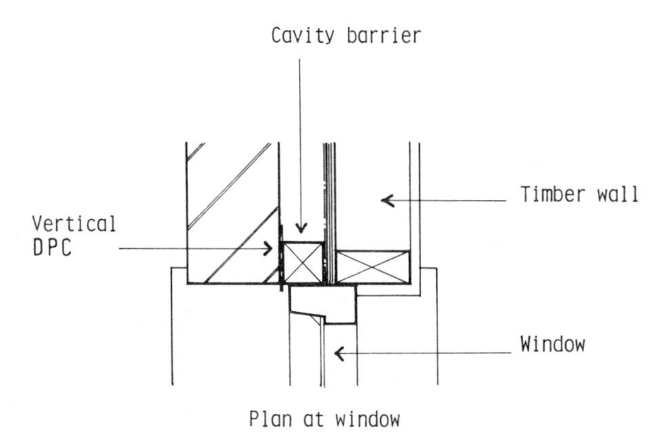

Fig. 5.5 Interpretation of Regulation C9(3)

(c) wind loads to CP 3, Chapter V, Part 2:1972 and the probability factor of S_3 must always be taken as 1 (see Chapter 6 on the assessment of wind pressure).

5.5.3 Regulation D8 – Structure above foundations

This is, of course, a most important part of the regulations written primarily for the designer of the building as a whole and hence aimed at the person or organization responsible for the stability of the building. However, notwithstanding this, all designers have a responsibility to ensure that any element they are dealing with, on completion, is capable of withstanding all loads imposed upon it and is able to be sensibly incorporated into the building as a whole. Above all, it has to be safe against failure.

The materials to be used in any design must always be examined against factors such as durability, life of the building, usage, maintenance, possible change of use, etc. In short, it is not enough to design a structural component for the loads alone if there are other factors which may have some effect upon the end result. With timber, for example, it is known that strength is reduced by moisture intake and, therefore, a designer would not proceed if he felt that possible damp penetration or exposure would impair the stability of an element.

5.5.4 Regulation D12

Regulation D12 gives the deemed-to-satisfy provisions for structural work of timber. Basically, three documents are referred to and these are:

CP 112, Part 2:1971
Schedule 6
CP 112, Part 3:1973

CP 112, Part 2:1971 is the current code of practice covering the structural use of timber. It is *the* most important Code of Practice which every designer must possess because it covers all permissible stress levels for most species of timber together with modification factors, plywood information, information on nails, screws, bolts, mechanical connectors and glue; information on workmanship and a section on test loading of structures. While touching upon testing, the designer should note that testing can only be entered into with the agreement of both parties concerned. In other words, in areas of doubt it does not necessarily follow that a checking authority has to accept testing.

Schedule 6 is a schedule of permissible span tables contained within the Building Regulations which relate to the use of floor joists, ceiling joists, joists for flat roofs, purlins, jack rafters and floor boarding. When reference is made to this schedule, no further justification by calculation is required. Care should be taken to read paragraphs 1 and 2 covering various conditions which must be complied with if the tables are to be used and also the species of timber must comply with the table provided. It is not uncommon for a designer to make no reference in his calculations to those members on the working drawings which have been obtained from Schedule 6. However, it is always sensible to be as helpful as possible and so a simple reference to Schedule 6 in the calculations where appropriate ties up what otherwise would be a loose end.

CP 112, Part 3:1973 is a specialized code dealing with the use of trussed rafters for roofs of dwellings. Again, for the specialist designer and supplier of timber roof trusses, this code of practice is a must and is as equally important as Part 2. Normally, this code of practice will be supplemented with an Agrément Certificate when the joints of the trusses are obtained by using specialist techniques, such as integral tooth plate connectors. Generally, for contractual reasons, it is prudent to leave the design of these joints to the system suppliers of the plates but there is no reason why a certificate should not be obtained and the joints designed accordingly. If this is to be done, then compliance with the International Truss Plate Association's joint design recommendations would be the right course to follow. A word of warning – if a certificate is obtained, ensure that the limiting date of the certification is still valid. If the date has run out, check with the Agrément Board that there have been no significant changes affecting the design capacities before making use of the information.

5.5.5 Regulations D17 and D18

These regulations cover the requirements for the localization of structural failure in the buildings of five or more storeys. As it is recommended that buildings constructed from timber should not exceed four storeys in height, these regulations will not apply.

5.6 Part E – Safety in fire

5.6.1 General

As far as this part of the regulation is concerned, when timber is used, it must always be considered to be combustible. Therefore, timber is nearly always protected by some non-combustible material in order to satisfy the fire regulations. The most common material, in this respect, is plasterboard which is used extensively in linings to walls, ceilings to floors and ceilings to roofs, with the total thickness varying to suit the period of resistance required.

5.6.2 Treatment

Timber can be treated with a chemical fire-retardant which will inhibit the spread of flame; however, such methods of treatment should be viewed with extreme caution for two principal reasons:

(a) fire-retardants tend to make the timber more hydroscopic than normal, hence increasing its tendency to absorb moisture and thus suffer a possible reduction in strength.
(b) fire-retardants can be corrosive and so will attack most ferrous fixings. The rate of corrosion, in some cases, is quite alarming.

5.6.3 Timber as a fire protection

Having said that timber is combustible, one should not overlook the fact that it is in itself a good fire protection material. The behaviour of timber is predictable with regard to rate of charring and loss of strength. Its state does not change rapidly as, for example, does steel and it has very low coefficients of thermal expansion. In addition, because timber has low thermal conductivity, the timber beneath the charred layer does not lose strength to any significant degree.

The property of sacrificial fire resistance is something which many designers overlook and it can come as a surprise to many, when the proposal for using timber as a fire resisting cladding is put forward. However, when considered carefully, it can be seen that in certain circumstances it could have some merit. For example, in protecting other materials, it could be used as a feature to enhance appearance.

5.6.4 The strength of timber in fire

BS 5268:Part 4, Section 4.1 was issued in 1978 and is entitled 'Fire resistance of timber structures – method of calculating fire resistance of timber members'. The publication of this code enables checks to be made on the performance of members in bending, in tension, in compression and a combination of bending + compression and bending + tension, where a known period of time resistance is required.

An example of how this code is applied is given in Chapter 11 (Section 11.3).

6

The assessment of wind pressure

6.1 Introduction

Wind pressure is probably the least understood and most ignored subject in structural element design. While all large structures will contain some degree of wind analysis in the calculations, many smaller ones do not and, consequently, the effects of wind can be overlooked in localized areas where it could adversely affect elements such as beams, columns and walls.

No matter the size of the structure, wind must always be considered, if only in reaching the conclusion that it may be ignored! When making this consideration, the designer should always bear in mind not only the overall size of the building but also the components which, when put together, form the whole. With buildings of predominantly timber construction, where the structural frame is considerably lighter in weight than the more traditional, wind may well be a critical factor, particularly in anchorage fixings. The British Standard Code of Practice to which reference must be made is CP 3, Chapter V, Part 2:1972 'Wind loading' and the prudent designer will make a point of reading this most carefully in order to obtain the minimum essential background knowledge.

Most designers will at some time have seen the damage caused by wind action on buildings manifested in the form of partial or total failures. Numerous partial failures take place practically every day and, certainly, every time high or unusual wind speeds are recorded. Probably the most commonly known damage which can be classified as partial failure occurs in roof coverings affected by wind suction; the commonest feature being the loss of tiles at ridges, eaves and verges where localized high pressures tend to concentrate. Another common feature in lightweight timber frame construction is sideways movement (sway) or buckling in individual members, both of which may be attributed to lack of lateral stability. Finally, one may occasionally witness total collapse for which there may be many reasons but whose cause can be attributed to wind action.

6.2 Background

The first British Standard Code of Practice covering wind loads was issued in 1944 giving specified basic wind speeds for various parts of the country. These speeds ranged from 55 mph to 85 mph and the unit pressure ranged from 7.0 lbf/ft^2 at ground level to 43.5 lbf/ft^2 at 200 ft above ground level. In 1952 the code was revised with four categories of exposure to the wind being introduced based on maximum wind speeds, averaged over one minute, of 45, 54, 63 and 72 mph.

Both codes referred to internal pressure, windward pressure and leeward suction with pressures on various roof and wall surfaces being obtained by applying specified coefficients to the basic wind pressure. The complexity of the subject is well known and, even today, knowledge is increasing as more and more studies are made of wind damage and more use is made of wind model analysis. So it is not surprising that the codes of 1944 and 1952 were sketchy and lacked real guidance born of analytical detail.

A major step forward in the information made available to the designer was made in 1972 with the introduction of CP 3, Chapter V, Part 2, 'Wind loads' and currently this code is still in use. However, we must accept that, in future years, this too will be replaced as our knowledge of the subject increases.

The significant achievement of the 1972 addition was the recognition of the effect of gust action and the important part that it played in causing damage to buildings. Wind is the movement of air through the atmosphere resulting from differences in air pressure caused by differential atmospheric heating over different parts of the earth. In the larger view, many local land influences affect the general wind flow and, at the other end of the scale, small and sometimes severe eddies may be due to wind passing a building or minor obstructions. These eddies are varied both in duration and complexity with the result that the wind speed varies both in position and time.

From meteorological studies and computer stored records, much statistical information was made available and from this was produced the map in the code which indicates the basic wind speeds called V. This is defined as the maximum 3-seconds gust speed that is likely to be exceeded, on the average, only once in 50 years at 10 m above the ground in open level country.

6.3 Factors S_1, S_2 and S_3

Once the basic wind speed has been selected for the design of the building, it must be modified by applying three factors to arrive at the design wind speed known as V_s. Briefly, these factors may be defined as follows:

(a) S_1 is a topographical factor and is used to allow for the acceleration of wind speeds found near exposed hill tops, the shelter afforded by deep valleys and also the funnelling effect of valleys.

If the building is close to a cliff top then Appendix D of the code may be used but not in combination with an increase in the topography factor.

(b) S_2 is a factor which takes into account variations of ground roughness around the structure concerned, variations of the wind speed averaging time, and variation of these wind speeds with height above ground level. The code measures this factor against three classes A, B and C which are defined as:

Class A: All units of cladding, glazing and roofing and their immediate fixings and individual members of unclad structures.

Class B: All buildings and structures where neither the greatest horizontal dimension nor the greatest vertical dimension exceeds 50 m.

Class C: All buildings and structures whose greatest horizontal dimension or greatest vertical dimension exceeds 50 m.

(c) S_3 is a statistical factor based on the probability of a severe storm of exceptional violence occurring over a period of exposure for the building measured in years.

Normally S_3 is taken as 1 and, in no case, must the period of exposure be taken as less than 2 years.

Sometimes S_3 can exceed 1 as in the case of a longer period of exposure than 50 years or where greater safety than normal is required.

6.4 Dynamic pressure (q)

The dynamic wind pressure (q) is defined in the code as $q = kV_s^2$ where k is 0.613 kg/m^3 and k is equivalent to half the air density at sea level, at a temperature of 15°C and at the standard atmospheric pressure of 1013.25 mbar and V_s is the design wind speed.

From this equation, it is readily observed that once the value of V_s has been defined then the all-important dynamic wind pressure may be found from the general application of the formula:

$$q = 0.613\, V_s^2\ \text{N/m}^2$$

Example of how to determine q

Given that

$V = 46$ m/s

$H = 5.5$ m

$S_1 = 1.0$

$S_3 = 1.0$

find the value of q.

Referring to Table 3 of the code for most timber design, Class B is used and so in this example with $H = 5.5$ m, S_2 is found to be 0.754 for Class B, Group 2. Now

$$V_s = V \times S_1 \times S_2 \times S_3$$

$$= 46 \times 1.0 \times 0.754 \times 1.0$$

$$= 34.70 \text{ m/s}$$

Using

$$q = 0.613\, V_s^2 \text{ N/m}^2$$

we have

$$q = 0.613 \times 34.7^2$$

$$= 738 \text{ N/m}^2$$

6.5 Pressure coefficients

Measurement of pressure under appropriate wind flow conditions is the only practicable way to determine them and so our knowledge of wind pressure is arrived at from the study of models in tunnels which simulate wind flow. Such tests take into account velocity gradient, turbulence and permeability; therefore, it is from visual studies and pressure measurements that our present understanding of the wind effects on buildings is derived. Figure 6.1 indicates the typical pattern of wind flow past a rectangular building with the wind blowing from West to East. From Fig. 6.1 it is observed that wind pressure

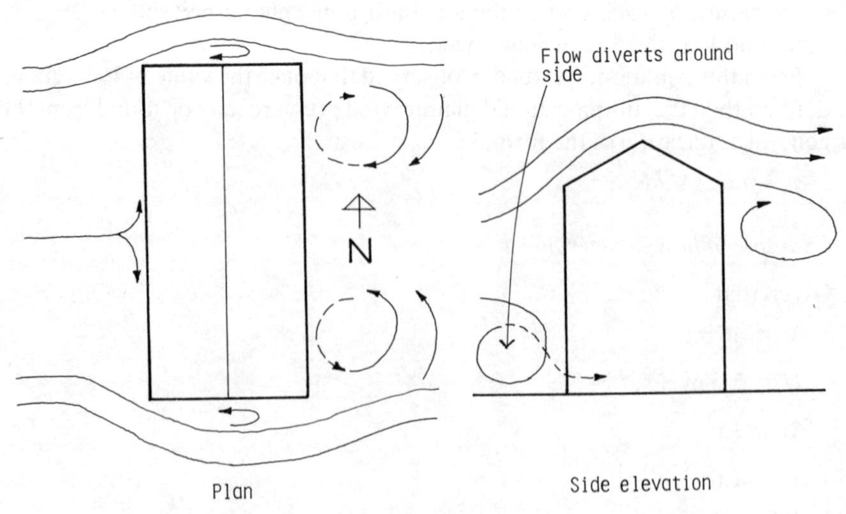

Fig. 6.1

($+V_e$) is built up on the windward side whilst suctional pressure ($-V_e$) occurs to the leeward side. Suction may also occur on the windward slope of the roof depending on whether or not the wind flow separates from the roof surface. This depends upon the pitch of the roof, with the wind separating more readily from a low pitch than from a steep one. Separation also increases with the angle of elevation of the flow over the edge of the roof, hence the influence of height of the building. From such studies as these, pressure coefficients are derived and this is expressed as:

$$\text{pressure coefficient } C_p = \frac{\text{pressure at any point on a surface}}{\text{dynamic pressure } q}$$

giving C_p+ (pressure) and C_p- (suction).

Other factors which will affect the wind flow and with it the pressure but which will not be discussed here are architectural features such as canopies, wind blowing at an angle to the building, L-shaped plans, complex building shapes and close grouping.

There are two kinds of pressure coefficients which must be considered and these are:

C_{pe} = external pressure coefficient

C_{pi} = internal pressure coefficient

If there is a dominant opening in a wall, say a large sliding door in a warehouse, then internal pressures can occur which will add significantly to the total pressure on walls. Appendix E of CP 3, Chapter V gives guidance on how to determine internal wind pressures, but in the case of buildings which we are to study and which form the majority, it is sufficient to assume that no predominant opening will occur during a severe storm and to take the internal pressure as the more onerous of $C_{pi} = +0.2$ and -0.3.

6.5.1 Wall coefficients for $C_{pe} + C_{pi}$

The code gives external pressure coefficients for walls of rectangular clad buildings based on ratios of eaves height (h) to the lesser horizontal dimension (w) and ratios of the greater horizontal dimension (l) to once again (w). These coefficients are set out in Table 7 of the code and they consider the four faces of the building with the wind blowing at angles to the faces of 0° and 90°.

Consider the rectangular building given in Fig. 6.2 and let us determine how we arrive at the external pressure coefficients for the walls. We must first evaluate the appropriate ratios in order that we may arrive at the correct section in the table to be used. Building height ratio:

$$\frac{h}{w} = \frac{8.0}{7.5} = 1.07 \qquad \text{which is} > \frac{1}{2} \text{ but} < 2$$

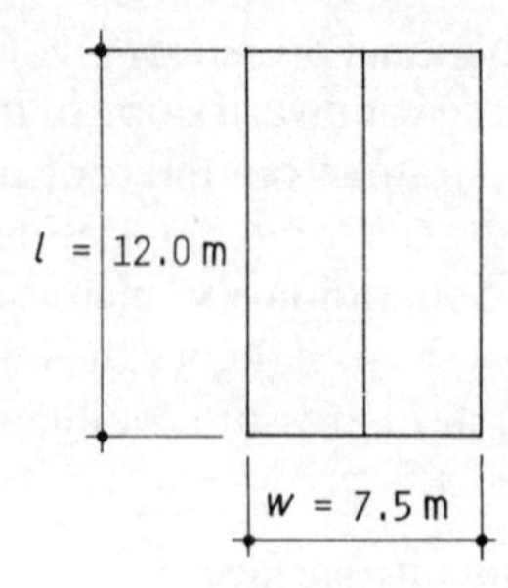

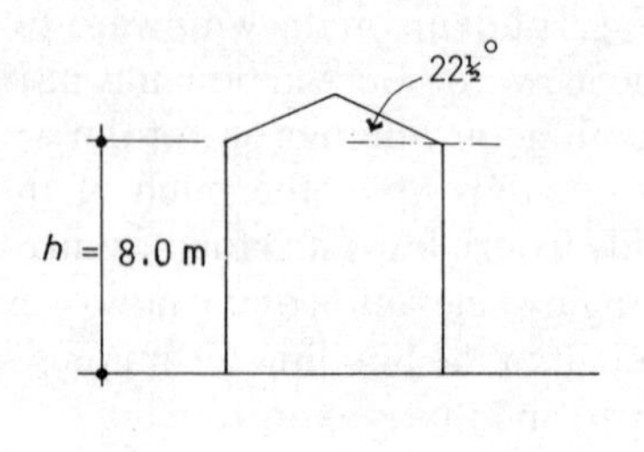

Fig. 6.2

Therefore in Table 7 we will first refer to the column containing the relationship of

$$2 < \frac{h}{w} \leq \frac{3}{2}$$

We must now examine the plan ratio in order to eliminate one of the two remaining relationships in this part of the table. Building plan ratio:

$$\frac{l}{w} = \frac{12.0}{7.5} = 1.6 \qquad \text{which is} > 1.0 \text{ but} < 4$$

and so the final relationship will be

$$\frac{3}{2} < \frac{l}{w} < 4$$

Now by reference to Table 7 and extracting the coefficients for the two angles and four faces Table 6.1 and Fig. 6.3 will serve to illustrate what would be shown in the calculations using

$$2 < \frac{h}{w} \leq \frac{3}{2} \quad \text{and} \quad \frac{3}{2} < \frac{l}{w} < 4$$

If we now assume that $C_{pi} = +0.2$ and -0.3 it is possible to arrive at the maximum total pressure coefficient on any given wall by subtracting these values algebraically. For this example the maximum external pressure is $+0.7$ and the maximum internal suction is -0.3 giving $C_{pe} + C_{pi} = +0.7 - (-0.3) = +1.0$. This total coefficient would now be used in the design of the

Table 6.1

α°	A	B	C	D	Local
0	+0.7	−0.3	−0.7	−0.7	−1.1
90	−0.5	−0.5	+0.7	−0.1	

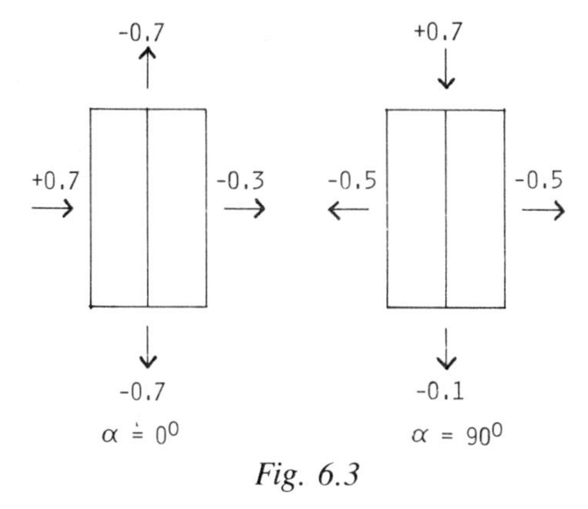

Fig. 6.3

wall elements. The localized suctional coefficient of −1.1 would normally be used to examine cladding designs and their associated fixings.

6.5.2 Roof coefficients for $C_{pe} + C_{pi}$

As previously stated, the flow of wind over the roof can either create suction or pressure depending on height to eaves level and depending on the amount of slope to the roof. Table 8 of the code gives the external pressure coefficients for pitched roofs of rectangular clad buildings for various ratios of h/w and for roof slopes of between 0° and 60°.

It will be seen that this table divides the plan area into eight equal parts which are considered to be symmetrical about the centre line of the longer side. These are designated E, F, G and H with localized eaves, verge and ridge widths of the value y, where $y = h$ or $0.15\,w$ whichever is the lesser.

Considering once again Fig. 6.2 with a roof slope of 22½° and a height to eaves of 8.0 m we have first the ratio

$$\frac{h}{w} = \frac{8.0}{7.5} = 1.07 \qquad \text{which is} > \frac{1}{2} \text{ but } < \frac{3}{2}$$

Therefore we must refer to that part of the table allotted to the relationship of

$$\frac{1}{2} < \frac{h}{w} \leq \frac{3}{2}.$$

By interpolation between slopes we can arrive at the following coefficients extracted from Table 8:

Roof angle	*α = 0°*		*α = 90°*	
	EF	*GH*	*EG*	*FH*
22½°	−0.575	−0.500	−0.800	−0.650

Note that the local coefficients are not indicated because these would normally only be used for design of claddings plus their fixings.

Once again we will consider the part played by the inclusion of the internal coefficients of C_{pi} = +0.2 and −0.3 leading to final loading diagrams indicated in Figs. 6.4 and 6.5

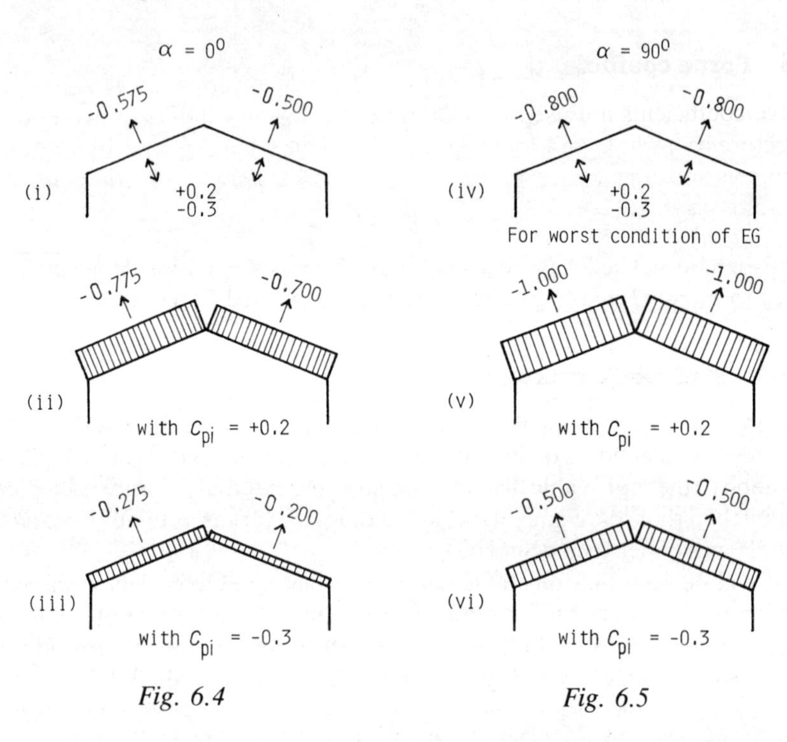

Fig. 6.4

Fig. 6.5

Depending upon the complexity of the design, it may not always be necessary to draw all of these diagrams but in evaluating the need the following design conditions should be considered:

(a) Uplift at roof seatings	Dead – wind condition (V)
(b) Reversal of roof member designs	Dead – wind conditions (ii) or (V) depending on span direction
(c) Design of long walls (*l*)	Dead + Super – wind condition (iii)
(d) Design of short walls (*w*)	Dead + Super – wind condition (vi)

By using conditions (iii) and (vi) in the design of the walls we ensure that we have established the maximum vertical load for the condition of bending (wind) + compression (D + S + W).

Before closing on the subject of pressure coefficients the reader's attention is drawn to Table 9 of the code which gives C_{pe} values for monopitch roofs. At a wind angle of $\alpha = 45°$ larger roof suctional coefficients are obtained than those for duopitch and therefore the designer must be aware of this phenomenon if dealing with designs which have monopitch roofs.

6.6 Force coefficients

Force coefficients are used to determine the overall wind loads acting on a structure as a whole and so they are primarily used in checking overturning moments and drag. Using pressure coefficients the force on any component of area A is given as

$$F = (C_{pe} - C_{pi})qA$$

whereas using a force coefficient C_f the formula would be

$$F = C_f q A_e$$

where A_e is the effective frontal area of building.

Logically, the algebraic sum of the pressure coefficients on the windward and leeward faces should equal the force coefficients given in Table 10 of the code. By comparison on particular designs, it will be found that this is not so and this can be explained by the results of test work. The work carried out on pressure coefficients covered a limited range of building proportions which could only justify the three height ratio groupings given in the tables. But the variation of C_f with plan shape and height/breadth ratio is better understood giving a better choice. In consequence, when assessing overall loading and turning about the base, the values for C_f in Table 10 should be used in preference to pressure coefficients in Table 7.

If we again take the example of Fig. 6.2 and consider the frontal elevation Fig. 6.6, the difference in the total coefficient causing overturning and loads on the foundations can be demonstrated.

For Table 10 and assuming a flat roof, we need

$$b = 12.0\,\text{m}, \qquad d = 7.5\,\text{m}$$

$$\frac{b}{d} = \frac{12.0}{7.5} = 1.60 < 2.0$$

$$\frac{l}{w} = \frac{12.0}{7.5} = 1.60 < 2.0$$

$$\frac{h}{w} = \frac{8.0}{7.5} = 1.07$$

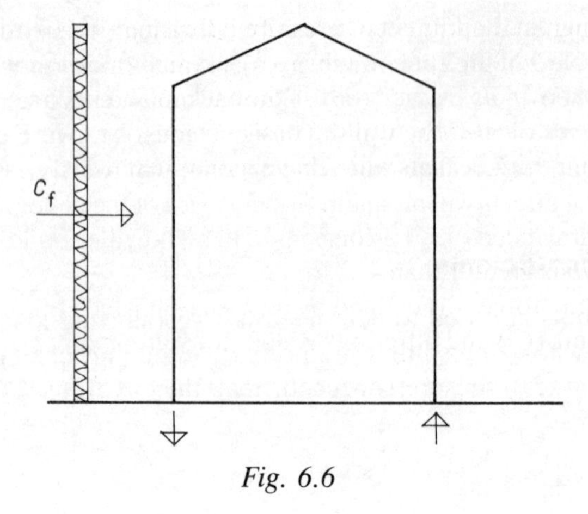

Fig. 6.6

By referring to the third column down in the table, given the above proportions, we can see that the value of C_f lies between 1.05 and 1.1 and so by interpolation

$$C_f = 1.05 + (1.1 - 1.05) \times \frac{0.07}{1.0} = 1.054$$

This may be compared with the value of 1.00 which was previously determined for $C_{pe} + C_{pi}$ which indicates that, for this exercise, there is an increase in loads at the base of around 5½% over and above what would have been determined had only the pressure coefficients been used. Similarly, the use of C_f gives a truer indication of the value of the factor of safety against overturning.

6.7 Frictional drag

Generally speaking, frictional drag is of no great significance in most buildings and it only becomes a consideration in structures of certain shapes or, in the

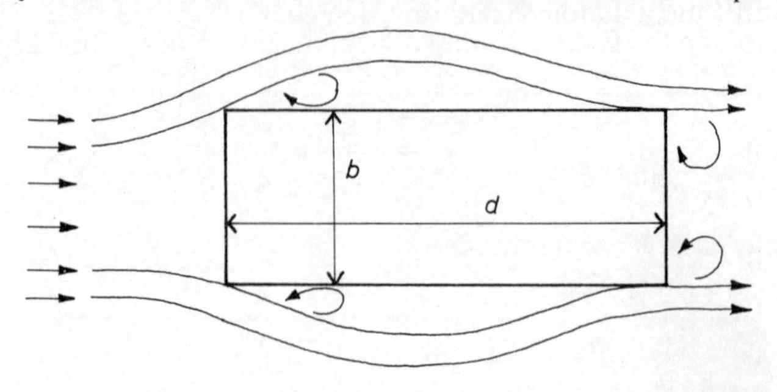

Fig. 6.7

case of rectangular buildings, where a building may be considered long in relation to its width. In long buildings, it is possible for the wind which has separated at the windward corners to flow back in and become re-attached at the leeward ends of the long walls. This is demonstrated in Fig. 6.7 where it will be seen that there is also some eddying at the windward ends back into the wall though the effects will be small. In Clause 7.4 of the code, the assumption is made that drag need only be considered in rectangular buildings where the ratio d/h or d/b is greater than 4.

The total drag force F is defined by an equation which allows for differing drag coefficients (C_f') and differing q values for walls and roof. If

$$h \leq b, \qquad F' = C_f'qb(d-4h) + C_f'q2h(d-4h)$$

or if

$$h \geq b, \qquad F' = C_f'qb(d-4b) + C_f'q2h(d-4b)$$

In each case, the first term in the equation gives the drag on the roof and the second the drag on the walls.

The code gives the following values for drag coefficients for different surface finish conditions:

$C_f' = 0.01$ for smooth surfaces without corrugations or ribs across the wind direction
$C_f' = 0.02$ for surfaces with corrugations across the wind direction
$C_f' = 0.04$ for surfaces with ribs across the wind direction.

Obviously when choosing one or more of these drag coefficients, some judgement must be made as to the classification of the surface finish but, for the main specifications of brickwork and concrete interlocking tiles, the coefficients should generally be taken as 0.01 and 0.02, respectively.

Example 6.1

The rectangular building indicated in Fig. 6.8 is clad with brickwork and contains concrete interlocking roof tiles. The dynamic wind pressure $q = 700$ N/m^2. Determine the frictional wind drag in the length of the building.

Now

$$\frac{d}{h} = \frac{50}{5.5} = 9.1 > 4$$

hence drag applies. $h < b$ therefore use

$$F' = C_f'qb(d-4h) + C_f'q2h(d-4h)$$

$$= 0.02 \times 700 \times 10(50 - 4 \times 5.5) + 0.01 \times 700 \times 2 \times 5.5(50 - 4 \times 5.5)$$

$$= 3920 + 2156 = 6076 \text{ N}$$

The magnitude of this figure could well be a consideration if the wind racking forces, as is likely, were to be resisted by specially designed panels. It must of course be added to the wind force generated through the wind blowing on the width b.

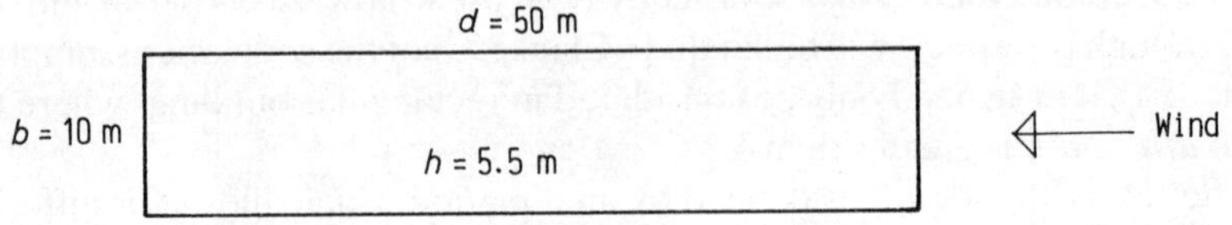

Fig. 6.8

6.8 Complete worked example

In order to summarize the contents of this chapter, we will examine a simple rectangular building (Fig. 6.9) in relation to the long side only and determine the factors of q, $C_{pe} + C_{pi}$, and C_f. These will be examined in regard to the need for holding down at the roof level the loads required for the wall design and the forces imposed at the foundation level. The building is assumed to be situated in open country with scattered windbreaks and of portal frame construction.

Given values are: $V = 48$ m/s, $S_1 = 1.0$, $S_3 = 1.0$, $l = 12.0$ m.

Step 1

Determine value of dynamic wind load q. Referring to CP 3, Chapter V, Part 2:1972 Table 3 and using Group (2) Class B with $H = 6.5$ m, then by interpolation $S_2 = 0.754$. Now

$$V_s = V \times S_1 \times S_2 \times S_3$$

$$= 48 \times 1.0 \times 0.754 \times 1.0 = 36.19 \text{ m/s}$$

Substituting for q

$$= 0.613 V_s^2$$

$$= 0.613 \times 36.19^2 = 803 \text{ N/m}^2$$

Step 2

Determine the $C_{pe} + C_{pi}$ values for roof and walls and draw the pressure diagrams.

Roof (Table 8 of the code): Considering only $\alpha = 0°$ with

$$\frac{h}{w} = \frac{5.5}{9.0} = 0.61 > \frac{1}{2}$$

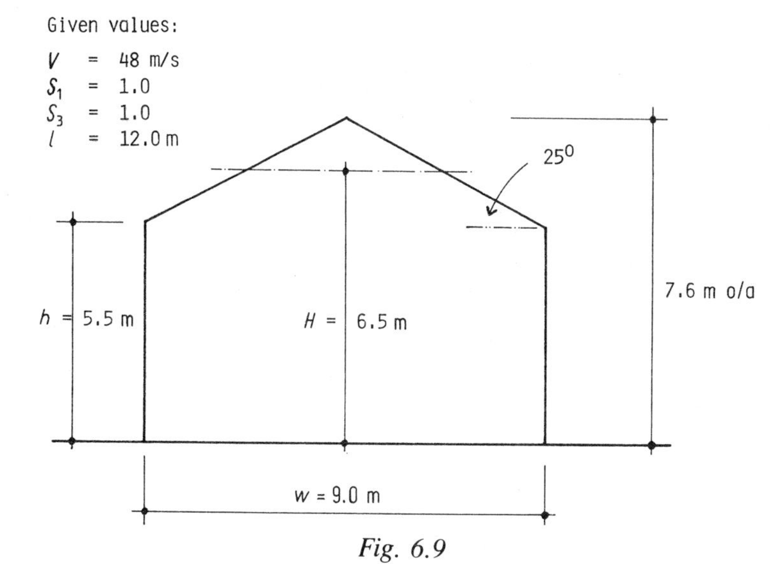

Fig. 6.9

therefore use

$$\frac{1}{2} < \frac{h}{w} \leq \frac{3}{2}$$

With roof pitch at 25° by interpolation we have:

	EF	GH
C_{pe}	$= -0.45$	-0.5
Use C_{pi}	$= +0.2$	and -0.3

leading to the pressure coefficients indicated in Fig. 6.10.

Walls (Table 7 of the code). With

$$\frac{h}{w} = 0.61 \text{ and } \frac{l}{w} = \frac{12.0}{9.0} = 1.33$$

From table use

$$2 < \frac{h}{w} \leq \frac{3}{2} \text{and } 1 < \frac{l}{w} \leq \frac{3}{2}$$

this gives:

α	A	B	C	D
0°	+0.70	−0.25	−0.6	−0.6

and these are transposed on to Fig. 6.11.

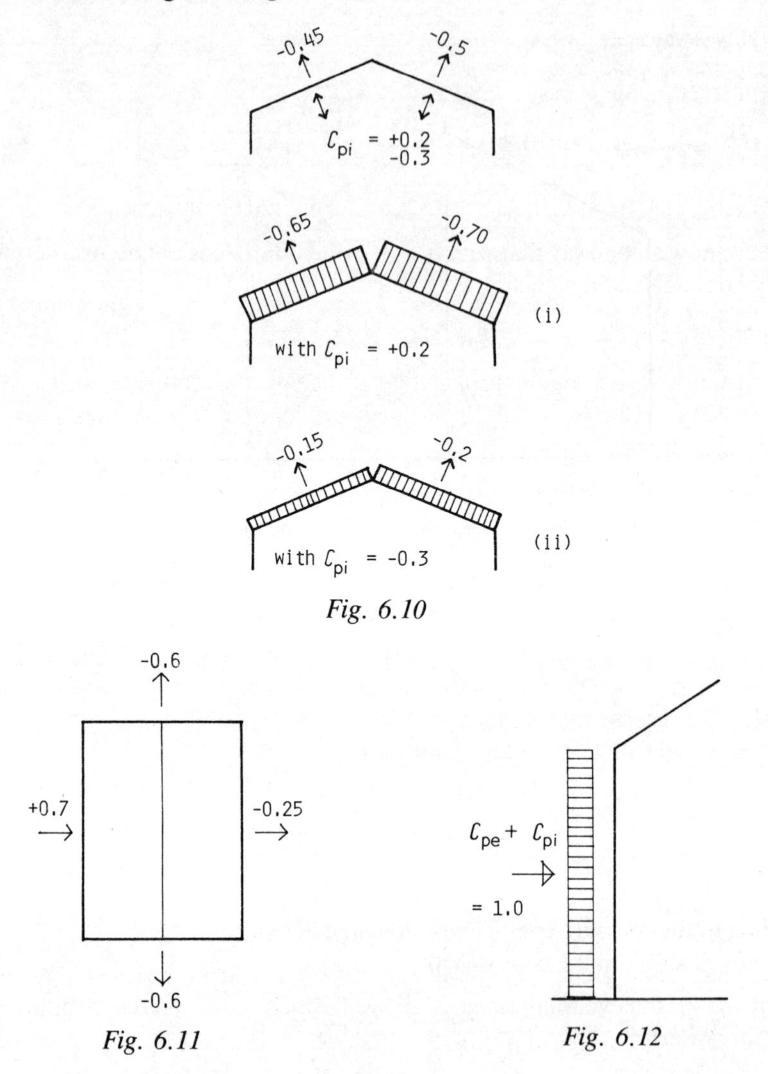

Fig. 6.10

Fig. 6.11

Fig. 6.12

Therefore by subtracting the pressure coefficients algebraically we have a maximum wall pressure coefficient (Fig. 6.12) equal to

$$C_{pe} + C_{pi} = 0.7 - (-0.30) = 1.0$$

Step 3

Determine if uplift occurs at roof seatings. The roof dead load will be taken as interlocking concrete tiles, battens, felt and joists giving:

Total dead load on slope = 0.685

Subtracting the average wind uplift (i)

$$= 0.803 \times \frac{(-0.65 - 0.70)}{2} = -0.542$$

Giving a net dead load = 0.143 kN/m^2

We are now able to say that with $\alpha = 0°$ wind uplift does not occur, therefore only nominal location anchorage is required.

Step 4

Determine the maximum vertical load at the eaves level for the wall designs and the maximum horizontal wind load for bending. Net unit area loading build-up is listed as follows:

Super load (on plan) = 0.75

Dead load (on plan) 0.685/cos 25° = 0.76

Wind uplift (ii) (on plan) $-(0.15 + 0.20)/(2\cos 25°)$ = -0.19

Giving a net total load W = 1.32 kN/m^2 on plan

$$\text{Maximum vertical eaves load} = W \times \frac{\text{span}}{2}$$

$$= 1.32 \times 9.0/2 = 5.94 \text{ kN/m}$$

$$\text{Maximum horizontal wind load} = q(C_{pe} + C_{pi})$$

$$= 0.803 \times 1.0$$

$$= 0.803 \text{ kN/m}^2$$

The final loading diagram for the wall design is as indicated in Fig. 6.13 allowing for the eventual design of the wall which is subjected to bending and compression.

Step 5

Evaluate the horizontal and vertical wind loads acting on the foundations. From Table 10 of the code with

$$\frac{l}{w} = 1.33 \text{ and } \frac{b}{d} = \frac{12}{9} = 1.33$$

plus

$$\frac{h}{w} = 0.61, \text{ use } \frac{l}{w} \geq 2 \text{ and } \frac{b}{d} \geq 2$$

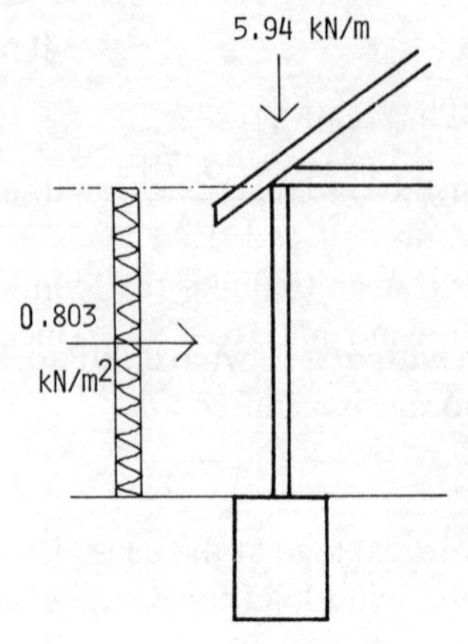

Fig. 6.13

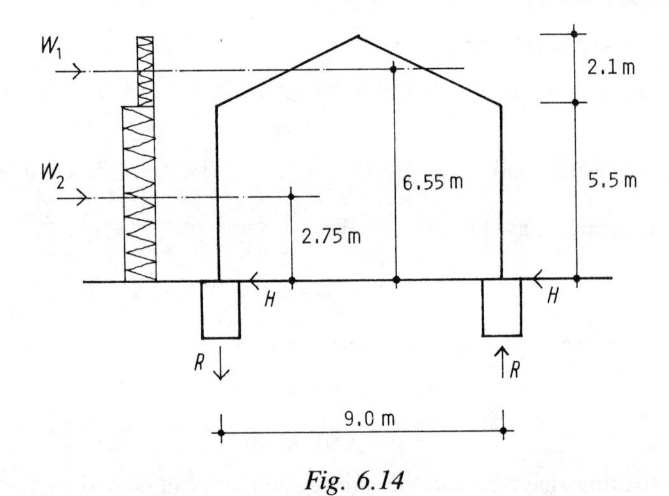

Fig. 6.14

by interpolation the value of $C_f = 1.011$. Now by superimposing the side wind loading on the cross-section, Fig. 6.14, we are able to determine wind reactions as follows:

$$W_1 = 0.803 \times 2.1 \times 1.011 = 1.71 \text{ kN/m}$$

$$W_2 = 0.803 \times 5.5 \times 1.011 = 4.47 \text{ kN/m}$$

$$\text{Wind overturning moment} = 1.71 \times 6.55 + 4.47 \times 2.75 = 23.50 \text{ kN/m/m}$$

Hence

$$R = \frac{23.5}{9.0} = \text{say } 2.61 \text{ kN/m}$$

and

$$H = (1.71 + 4.47)^{\frac{1}{2}} = 3.09 \text{ kN/m}.$$

In the final analysis the vertical wind reactions would summate to the vertical loads caused by the superimposed and dead loads to increase and reduce accordingly, the total loads transmitted into the foundations. Similar calculations would be required for wind angle $\alpha = 90°$ to include for wind drag.

See Case Study 8 (page 237) for a further example containing calculations for stability.

7

How to take off loading

7.1 Introduction

The assessment of applied loading is possibly the most important task of the designer, for, once done, it controls the whole solution to the problem. It follows, therefore, that particular care and attention are required at this early stage of the design in order to ensure that no fundamental mistakes are made. It is worth spending some time over this task and before proceeding to the next stage, at least one check should have been made on the loading figures obtained. For the elements of a structure which we are to design, the loadings should not present too difficult a problem and, given the correct information at the onset, solutions will easily follow.

It is necessary to understand that structural design related to building by its very nature is not an exact science and, while one should aim for accuracy at all times, a reasonable degree may be said to exist within the bounds of $\pm 5\%$. In order that the designer may determine loading within this reasonable degree of accuracy, the following sections should serve as a guide to good practice procedure.

7.2 Architectural information

Very often the architectural drawings supplied may be in a preliminary form. It is the task of the designer to determine that the architect has supplied sufficient information for the element size to be determined. Never fail to seek further and better information if it is thought necessary. If this cannot be given, then heavily qualify your solution by notes added to all particulars which you supply.

You need to determine all controlling dimensions related to plan and height and this sometimes requires careful scrutiny of the drawings. Generally speaking, architects do not provide dimensions of width and length on any cross-section and so the plan drawing should be examined for these. Floor to ceiling heights are normally given on cross-sections but, be careful, these dimensions very often relate to finishes, e.g. finished floor to finished ceiling. A common error to be found is non-correlation of similar dimensions; for example, a row of dimensions may be given between walls which pick up all openings but, when checked against the overall dimensions, a discrepancy is found.

Another common feature in architectural detailing is to stagger the section through the plan area of the building and so the designer may well be misled into believing that certain loading exists when, in fact, it does not. So always carefully follow the path of plan section marks and make your own adjustments as necessary. Finally, if you have to apply a scale to any drawing, then say so.

7.3 Specification of construction

Where architectural drawings are provided, the specification to most parts of the construction will be found on the drawing. If this is not the case, then information should be sought from the architect or designer of the building. Should a written specification be provided then this will take precedence over the drawing; though it is wise to compare the two and raise queries if discrepancies are discovered.

In some instances, the specification may be written in general terms, for example, the construction to a roof may read as follows:

> Interlocking concrete tiles on 25 × 50 battens on sarking felt on sarking on specialist roof trusses with insulation and 12.7 mm plasterboard.

This type of specification prompts many questions: What type of tiles? What pitch of battens? What type of sarking? What centres and slope of trusses? What thickness of insulation and is the ceiling dry-lined or skim finished? Where questions of this kind remain unanswered, then the designer is vulnerable to criticism. So the message is there, look for clear and concise specifications to all parts of the construction and never guess. Again, if information is not obtainable, say so and explain with concise notes what has been assumed in the design.

7.4 Determination of weights of materials

There is much literature on this subject available to the designer but the commonest reference is BS 648:1964 'Schedule of weights of building materials'. Where specialist materials are involved, then, normally, technical literature is available and densities will be found here. CP 112:Part 2:1971 gives the weights for all of the timber species occurring in the code and also most plywood. Weights are given in values of kg/m^3 and kg/m^2, the latter is always related to a specified thickness. As the designer will normally be working in units of kN, then the simple expedient of dividing by 100 will obtain the correct values.

7.5 Position of element in building

The position of the element within the building as a whole should next be considered and the drawings examined to determine all parts of the construction which impose load upon the element. For example, a lintel over a window may have loads from floor spans, various storey wall heights and roof; roof loading may also possess water tanks or service loadings which could impart additional loads. Careful reading of the drawings will give the designer an overall picture of the total construction and from this will be gleaned the part played by the element under consideration.

7.6 Plan area and storey heights

The element must now be examined in its particular locality in relation to position on plan and the number of storey heights which affect it.

Directions of spans of floors are most important so are positions and heights of partitions. A sketch or two at this stage is most useful and preferably to scale to give a sense of proportion. Be as accurate as possible in the dimensional co-ordination and always put a second check on all controlling dimensions. Do not rush your work and take time and care in the preparation of your sketches.

7.7 Classification of loads acting on an element

The geometry of the loading patterns will determine their classification but normally, once determined, they may be broken down into four basic classifications:

(a) PL points loads (or concentrated loads)
(b) UDL uniformly distributed loads (or partial UDL) – square or rectangular in geometric terms
(c) TL triangular loading
(d) Par L parabolic loading (this one is not as common as the first three).

Once the plan or elevational shapes have been determined, as in Section 7.6, their total load and point of application on the element may be determined. Again, a sketch to scale should be prepared. Having done this, the element is now ready for analysis and shears, bending moments and deflections may be calculated.

7.8 Special load conditions

There are many occasions when special loading will occur without it being made clear and so the designer must be alert to this possibility. Cold water storage tanks are one example of a special loading which commonly occurs and yet is overlooked in design, particularly domestic storage in roof spaces.

Service loadings such as heating and ventilating should be watched for especially in office development. The trunking in such cases is generally supported at large centre to centre spacing which can cause appreciable concentrated loads or partial UDLs. Where gymnasiums form part of the building, one should check for climbing ropes and other special equipment.

Sliding and roller shutter doors are another source of special loading, together with top hung folding partitions. The latter can often create higher bending stresses and deflections because the bunching of a partition can occur at the centre span of a beam, thereby simulating a concentrated load rather than a UDL. Similarly, the total shear load can apply predominantly at one support.

Valleys and north light construction can lead to a heavy build-up of snow or trapped water, but particularly the former. The maximum snow loading required in the UK is 0.75 kN/m^2 but it is very possible to exceed this loading in special roof construction shapes and the designer may well have to make an assessment.

The foregoing are some examples of special loading conditions and it makes good sense to develop an awareness of those types of construction which can generate such special loading.

8

Standard format for drafting calculations

8.1 Introduction

The drafting of structural calculations requires thought, care, attention to detail and neatness of presentation. There is nothing more irritating to a checker than poorly presented work which is both difficult to understand and almost impossible to decipher. This kind of presentation is bound to attract questions leading to misunderstanding and possible rejection. On the other hand, neatly presented calculations which have clearly been given a great deal of thought and attention rarely present the checker with problems. By examining a standard method of approach, it should be possible, with little variation, to adopt a procedure of presentation which will always be acceptable to the checker. The following sub-headings can be used as points of reference:

- Sheet layout (A4 size)
- References
- Concept
- General Loading
- Wind Loading (value of q)
- Element design and loading
  - (a) Timber species
  - (b) Basic stresses
  - (c) Design loading and philosophy, e.g. pinned, fixed, continuous
- Summary of findings
- Notes and sketches

and expanded upon as follows.

8.2 Sheet layout

The sheet should preferably be standard A4 size which will conform to the size adopted by the majority. The use of a 20 mm margin to the left-hand side will allow for punching for filing wherever required. Adjacent to this margin, there should be a column approximately 25–30 mm wide in which all chosen sub-titles can be written. Normally, to the top or bottom of the sheet will be printed the name of the firm preparing the calculation, together with suitable spaces for project titles, job and sheet number and date (see Fig. 8.1).

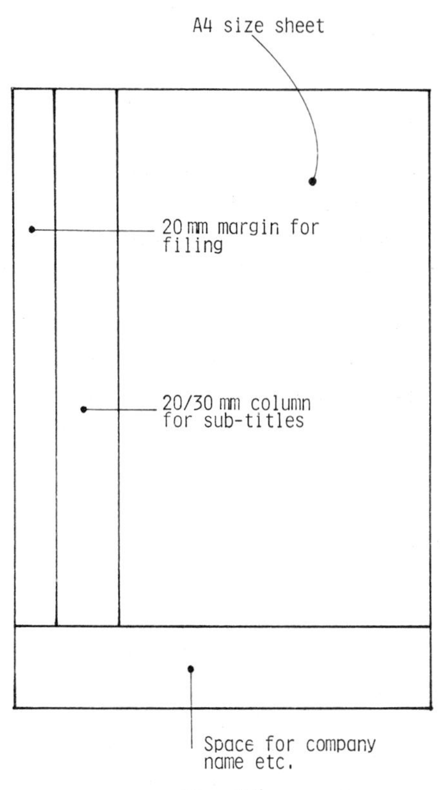

Fig. 8.1

8.3 References

Under this sub-title you simply draw the checker's attention to those documents which you need to refer to in order to complete your work. A simple example is British Standard Codes of Practice which should always be referred to without exception. Reference should also be made to any books, certification, research notes, specialist literature, etc. which are fundamental to the background and build-up of the calculations. This important sub-title gives the checker the opportunity of referring to the same literature and makes his job that much easier.

8.4 Concept

This amounts to a brief description of how the designer sees the structure working and the manner of approach adopted. It helps to give the overall picture to the checker and it is a good procedure for the designer to follow because it will also help to clarify his own thinking and eliminate ambiguity. Much time can be wasted before it becomes evident that checker's and designer's thinking are at cross-purposes.

8.5 General loading

It is always a good policy to set out the general loading in terms of unit area (kN/m^2) listing all the materials which form the particular section considered and giving a load for each material. Unit area loading will cover all floor levels, roof and external and internal walls. Each build-up will, of course, carry the appropriate superimposed loading.

In addition, under this heading should be listed any special loads which may apply, e.g. water tanks, plant and equipment, climbing ropes, folding doors and partitions, etc. (see Chapter 6). By following this logically through all levels of the structure, you will save much time in unnecessary repetition at later stages in your work.

8.6 Wind loading

Wind loading may not always form part of the work but where it does, then it is equally important to consider it carefully and to set it out immediately following the General Loading section. The object here is to determine the value of the dynamic wind loading (q).

If the element under consideration involves checking the effects of wind uplift, then immediately after finding q, the dead load (long term) should be evaluated and subtracted from the q multiplied by the appropriate $C_{pe} + C_{pi}$ factors, to ascertain if the net value is, in fact, an upward wind load. If so, then the calculations which follow will have to consider holding down anchorage. If not, then always state clearly 'Wind uplift does not occur' (refer to Chapter 6 on the assessment of wind pressure).

8.7 Element design and loading

This is where the designer, having now determined all of the general controlling loads, must complete the detailed design of the element or elements under consideration. The loading will normally be clarified with a simple line diagram and the design should indicate any philosophy assumed, i.e. pinned ends, fixed supports, continuity, etc. The timber species should be given with all basic stresses. If several elements are to be designed and the same stress graded material is to be used throughout, then to save repetition, this work could be given a specific sub-title and appear after 'Wind loading'.

8.8 Summary of findings

If several elements have been covered then a summary of sizes in a list form will help to draw the checker's attention to the designer's findings. For example:

Item	*Size*	*Grade*	*Species*
Post (Grid A14)	100 × 100 sawn	SS	Hem-Fir
Beam No. 21	2/37 × 145 actual	M75	Whitewood
Truss Type A	Top Chord 35 × 97	M75	Whitewood
	Btm Chord 35 × 120	M75	Whitewood
	Webs 35 × 72	M50	Whitewood

8.9 Notes and sketches

At all times during the process of producing the calculations, always remember that notes and sketches will help the checker in areas where there could be doubt. Sketches are particularly useful and can often say far more than any amount of words.

If a line diagram is drawn to say 1:100 scale or other, as necessary, it gives the right proportions to the problem and puts matters into their correct perspective. For example, consider the ply box beam free-hand sketch in Fig. 8.2 and compare this with a scaled detail in Fig. 8.3. It is readily observed that the scaled detail is more acceptable to the eye and reduces risk of errors occurring. The amount of additional time taken is more than compensated for by the improvement in accuracy.

Similarly, a note of explanation helps to guide the checker in regard to the designer's reasoning and he is not left with a guessing game. For example, the following ending to a deflection calculation could raise queries:

$$d_p = 0.003 \times 5200 = \underline{15.6}\ \text{mm} < 16.8\ \text{mm} \qquad \text{applied say OK.}$$

Why say OK? This will be an instinctive question by the checker; after all, the percentage increase on permissible is around 8% which would probably be argued is too high. For this particular case, let us suppose that the finish to the beam is dry lined and the greater part of the dead load is applied prior to fixing the dry lining and so the designer is justified in allowing a larger deflection, up to say 0.0045 of the span. A few words of explanation to this effect will save time by allowing the checking process to proceed smoothly without recourse to questions.

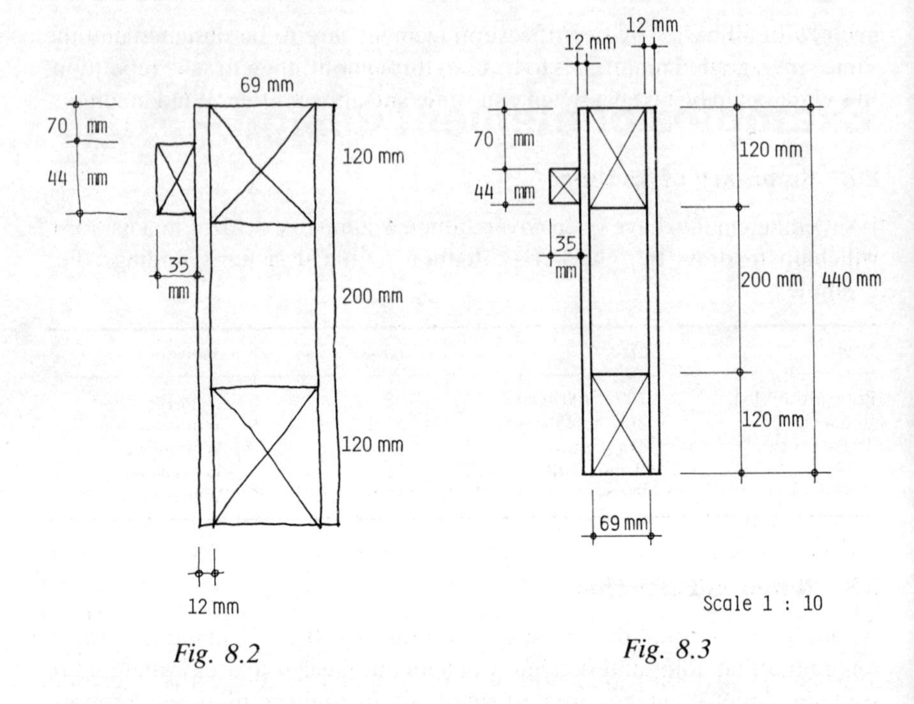

Fig. 8.2

Fig. 8.3

8.10 Conclusion

It is hoped that this chapter will have served to improve the student's understanding of the art of presenting calculations, for it is an art and one well worth particular attention and much practice. It is beholden to all structural designers always to present their particular skill in a manner which will attract approval rather than criticism.

9

Examples of element design – Part I

9.1 Introduction

The previous chapters will have given the reader an insight into those subjects which it is necessary to understand in order that an approach may be made to the design of most structural elements using timber as the medium of construction. It is, of course, possible for an experienced designer to arrive at element sizes by simply using his general knowledge of theory of structures and by referring to the appropriate codes of practice. However, without possessing experience of the general performance of the material chosen, serious errors can be made.

Having, for example, read the chapters on materials, physical properties and stress grading, the designer will be much better equipped to deal both safely and economically with the design of structures where timber is to be used. In the examples which follow, an attempt will be made to give step by step guidance through those processes which are necessary to arrive at reasoned solutions.

9.2 Solid section and glue laminated beams

In these studies we will consider beams to be simply supported but this is not to say that continuous timber beams may not be used. Continuity over supports is as permissible in the use of timber as it is in, say, concrete and steel, provided that the correct theory is used.

In all cases we must examine the three principal criteria of bending, shear and deflection and, to some extent, it may be necessary to check bearing stresses.

CP 112 refers to factors k_1, k_2, etc., which are various modification factors applied to the basic design stresses and which are used as required in the worked examples which follow.

It is logically argued by some designers that the effective span of a simply supported beam may be taken to be equal to the clear distance between the faces of the supports. This is because the required bearing area is generally small if expressed as a percentage of the span. However, it is probably more prudent to adopt the readily acceptable procedure of using centre to centre of bearings for the following reasons:

(a) This assumption would not be subject to question and possible tedious discussion with a checker.
(b) It could be argued that, in some cases, because deflection is related to a cube function of the span even a small increase in length could be significant.

To expand on point (b), if we take the case of a beam with a clear span $L_c = 3.0$ m and an effective span $L_e = 3.10$ m then the relationship is

$$\frac{L_e^3}{L_c^3} = \frac{3.10^3}{3.0^3} = 1.103$$

giving an increase in deflection in excess of 10% compared with a nominal linear increase of 3.3% when comparing the spans alone. Therefore, in the examples which follow, we will always use L_e = centre to centre of allotted bearings and, by working step by step, we will obtain a clear illustration of how the designs are accomplished.

Example 9.1

The loaded beam in Fig. 9.1 is one of four spaced at 600 mm centres forming part of a roof construction and the task is to check that the size is satisfactory.

Step 1 Determine the magnitude of the reactions and hence check for the maximum shear stress.

Step 2 Determine the point of zero shear and from this the maximum bending moment. Check the moment capacity of the given beam section.

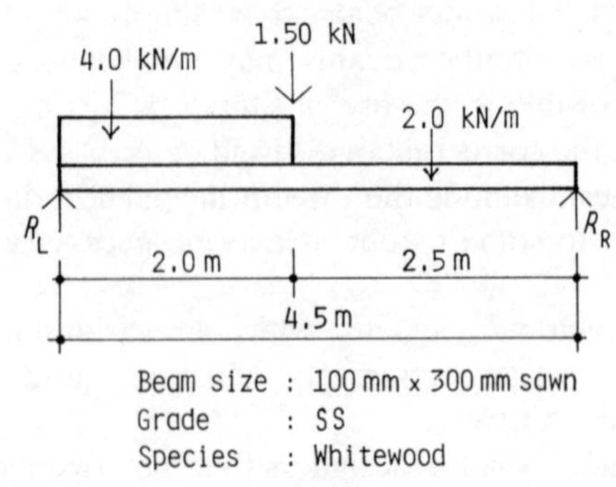

Fig. 9.1

Step 3 Check the maximum deflection against the permissible.

Step 1

To determine the magnitude of the reactions, moments will be taken about a line drawn through R_R:

$$4.5R_L - 1.50 \times 2.5 - 2.0 \times \frac{4.5^2}{2} - 4.0 \times 2.0 \times 3.5 = 0$$

$$4.5R_L - 3.75 - 20.25 - 28.0 = 0$$

$$4.5R_L - 52.0 = 0$$

$$R_L = 11.56 \text{ kN}$$

$$R_R = 1.50 + 9.00 + 8.00 - 11.56 = 6.94 \text{ kN}$$

Hence the maximum applied shear (V) = 11.56 kN (R_L). Now shear stress

$$v = \frac{3V}{2bd} = \frac{3 \times 11.56 \times 10^3}{2 \times 100 \times 300} = 0.578 \text{ N/mm}^2$$

and the permissible shear stress is

$$v_p = v_g \times K_{12} \times 1.1 = 0.70 \times 1.25 \times 1.1 = 0.96 \text{ N/mm}^2 > 0.578$$

where v_g is taken for the green condition (see later note on moment of resistance).

Step 2

Working from the left-hand reaction (R_L) the position of the zero shear, distance L_1 (Fig. 9.2) may be found

$$11.56 - 6.0 \times L_1 = 0$$

$$L_1 = \frac{11.56}{6.0} = 1.93 \text{ m}$$

Having now determined the point in the beam where the zero shear occurs, the maximum moment is found by taking moments about this point to one side

$$\text{Max. BM} = 11.56 \times 1.93 - 6.0 \times \frac{1.93^2}{2}$$

$$= 22.31 - 11.17$$

$$= 11.14 \text{ kN m}$$

Now that the external applied moment is known, we must know if the beam section provided is sufficiently large to generate an internal moment to resist the externally applied one.

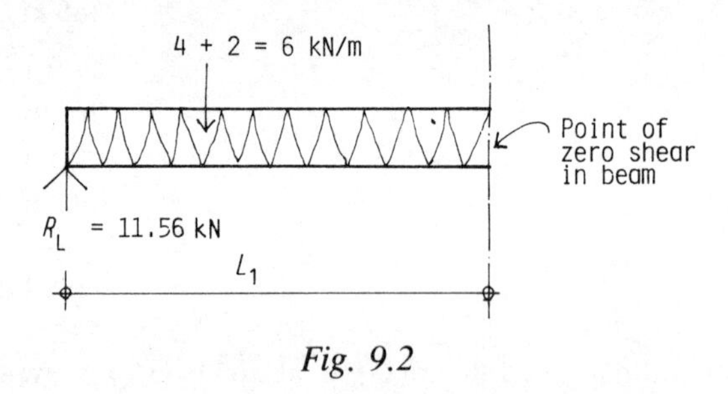

Fig. 9.2

The moment of resistance is expressed as:

$$\bar{M} = Zf_{\text{ppar}}$$

where

$$Z = \frac{bd^2}{6} = \frac{10.0 \times 30^2}{6} = 1500 \text{ cm}^3$$

For a section of this size, there is a possibility that the moisture content would be high with substantial shrinkage taking place; therefore, green stresses should be used. Referring to Table 3.6 for SS Whitewood:

$$f_{\text{ppar}} = f_{\text{gpar}} \times K_{12} \text{ (med. term for roof loading)} \times 1.1 \text{ (load sharing)}$$

$$= 5.9 \times 1.25 \times 1.1 \quad = 8.112 \text{ N/mm}^2$$

$$\bar{M} = \frac{1500 \times 10^3 \times 8.112}{10^6} = 12.17 \text{ kN m} > 11.14 \text{ kN m} \qquad \text{OK}$$

Step 3

The final step is to determine if the deflection is within permissible limits bearing in mind the guidance given in the code of practice (it may be assumed that the dead load does not exceed 60% of the total – Clause 3.13.2 of CP 112).

By referring to Appendix A we can arrive at an equivalent UDL by breaking down the total loading into three separate cases (Fig. 9.3) which, when added together, will give the approximate maximum deflection. These cases are:

(1) $PK_b = 1.50 \times 1.577 = 2.37$

(2) $WK_b = 8.0 \times 0.934 = 7.47$

(3) $WK_b = 9.0 \times 1.00 = 9.00$

$$W_e = 18.84 \text{ kN}$$

Using

$$d = \frac{5W_eL^3}{384EI}$$

$$d = \frac{5 \times 18\,840 \times 4500^3}{384EI}$$

where

$$E = E_{(mean)} = 8200 \text{ N/mm}^2 \text{ (green)}$$

and

$$I = \frac{10 \times 30^3}{12} = 22\,500 \text{ cm}^4$$

$$= \frac{5 \times 18\,840 \times 4500^3}{384 \times 8200 \times 22\,500 \times 10^4} = 12.12 \text{ mm}$$

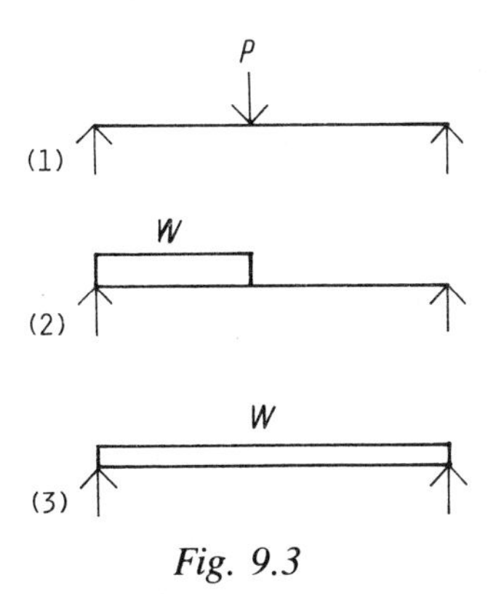

Fig. 9.3

Permissible (CP 112) deflection = $0.003 \times 4500 = 13.5$ mm and as this is greater than the applied deflection the beam may be said to be satisfactory.

Example 9.2

Determine the size of section required to form a simple beam in the floor of a house which trims a staircase opening. The loading to be carried by this beam is indicated in the line diagram shown in Fig. 9.4.

Step 1 Determine the value of I from the formula for the total deflection. Knowing I determine a size for the beam.

Step 2 Determine the maximum bending moment and so check the maximum bending stress.

Step 3 Determine the value of the maximum reaction and check the maximum shear stress.

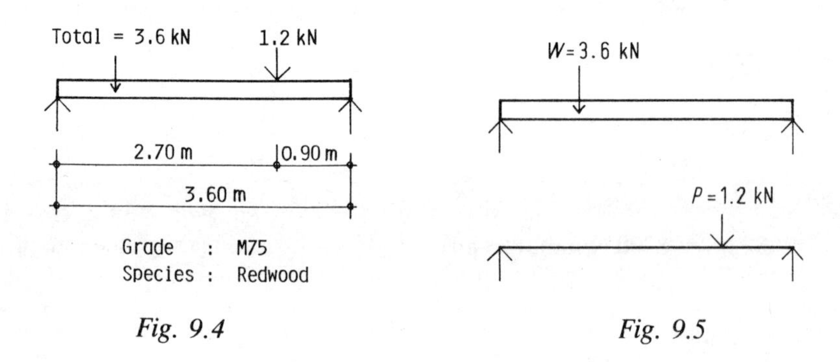

Fig. 9.4

Fig. 9.5

Step 1

By referring to Appendix A and breaking down the loading into two standards (Fig. 9.5), we obtain each of their equivalent loads which are added together and related to the value of a UDL deflection. From this simple equation may be found the value of I required for the section:

$$WK_b = 3.6 \times 1.0 = 3.60$$

$$PK_b = 1.2 \times 1.1 = 1.32$$

$$W_e = 4.92 \text{ kN}$$

$$\text{Total } d = \frac{5W_eL^3}{384EI}$$

$$d = \frac{5 \times 4920 \times 3600^3}{384EI}$$

Because we are dealing with a staircase trimmer, it constitutes an isolated beam and, therefore, the value for $E = E_{(min.)} = 6700$ N/mm^2 (Table 3.11) for Redwood and so, substituting into the above equation and simplifying, we have:

$$d = \frac{4.461 \times 10^8}{I}$$

Now by relating d to d_p for the permissible in the code of practice, we have:

$$d = d_p = 0.003 \times 3600 = 10.8 \text{ mm}$$

and so

$$10.8 = \frac{4.461 \times 10^8}{I}$$

giving

$$I = \frac{4.461 \times 10^8}{10.8} = 4131 \times 10^4 \text{ mm}^4$$

As the beam is forming part of a domestic floor, we will assume that a maximum joist depth of 200 mm applies and by substituting this value into the formula for I, we can arrive at a minimum dimension for the width of the beam:

$$I = \frac{bd^3}{12} \quad \text{with } d = 194 \text{ mm (finished depth)}$$

and

$$I = 4131 \times 10^4 \text{ mm}^4$$

transposing,

$$b = \frac{4131 \times 10^4 \times 12}{194^3}$$

$$= 68 \text{ mm}$$

Choose minimum beam size of 2 No. 38 × 200 mm (35 × 194 finished) M75 floor joists securely nailed together to form one unit. This gives a maximum overall width $b = 70 \text{ mm} > 68$ mm required. The reason for choosing 38 mm as the basis for the final size is because this is a readily available thickness graded to M75 standards.

Step 2

The approximate maximum bending moment may be taken as:

$$\text{Max. } M = \frac{WL}{8} + \frac{Pab}{L}$$

$$= \frac{3.6 \times 3.6}{8} + \frac{1.2 \times 0.90 \times 2.70}{3.6}$$

$$= 2.43 \text{ kN m}$$

giving a maximum bending stress of:

$$f_{\text{apar}} = \frac{M}{Z} \text{ with } Z = Z_x = \frac{bd^2}{6}$$

$$= 7.2 \times \frac{19.4^2}{6} = 451 \text{ cm}^3$$

Therefore

$$f_{\text{apar}} = \frac{2430}{451} = 5.40 \text{ N/mm}^2$$

$$f_{\text{ppar}} = f_{\text{gpar}}$$

because the beam is contained in a floor and so all loading is considered to be long term. In addition, it is a principal member and therefore load sharing does not apply.

f_{gpar} (Table 3.11) $= 10.0 \text{ N/mm}^2 > 5.40$ applied Satisfactory

Step 3

The maximum reaction is

$$\frac{W}{2} + \frac{Pb}{L}$$

$$= \frac{3.6}{2} + 1.2 \times \frac{2.7}{3.6} = 2.70 \text{ kN}$$

giving a maximum shear stress of

$$v = \frac{3V}{2bd} = \frac{3 \times 2700}{2 \times 72 \times 194} = 0.29 \text{ N/mm}^2$$

permissible $v_{\text{p}} = 1.28 \text{ N/mm}^2 > 0.29$ applied Satisfactory

In the previous two examples either $E_{min.}$ or E_{mean} is used depending upon the designer's interpretation of Clause 3.13.2 of CP 112:Part 2:1971. The code requires the use of $E_{min.}$ when a member is isolated but when four or more members can be considered to share a common load E_{mean} may be used.

The various grade stress tables in the code give values for $E_{min.}$ and E_{mean} only but when members up to a maximum of four act together, it is reasonable to use a value of E_N where N is the number of pieces considered.

The formula for deriving E_N may be simplified to read:

$$E_N = E_{mean} - \frac{(E_{mean} - E_{min.})}{\sqrt{N}}$$

For example, if a beam is made up of three floor joists bolted together and supporting a common load then for the case of European Whitewood visually stress graded to Table 3.2:

$$E_{mean} = 8300\ \text{N/mm}^2 \text{ and } E_{min.} = 4500\ \text{N/mm}^2$$

$$E_3 = 8300 - \frac{(8300 - 4500)}{\sqrt{3}}$$

$$= 6106\ \text{N/mm}^2$$

The advantages in such a case over the use of $E_{min.}$ are obvious. In Chapter 3 we discussed the method used for stress grading glue-laminated construction and in the example which follows we will discover how CP 112:Part 2, Clause 3.7 is used to derive the size of a horizontally glue-laminated beam section.

Example 9.3

A glue-laminated beam indicated in Fig. 9.6 is required to span a clear opening of 6.0 m with a bearing each side of 100 mm. It carries a uniformly distributed total roof load of 17 kN and must not exceed 90 mm in width. Determine the beam's depth and the number and size of laminates to be specified assuming the compression edge to be held in line by incoming purlins.

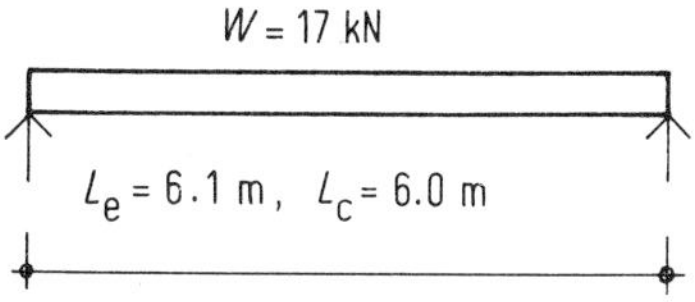

$W = 17$ kN

$l_e = 6.1$ m, $l_c = 6.0$ m

Fig. 9.6

Step 1 Determine the bending moment and shear.
Step 2 Choose a laminate thickness and a beam depth plus the grade to be used.
Step 3 Examine and determine all modification factors.
Step 4 Evaluate the bending and shear stresses and hence the beam's strength and determine the total deflection. Check the bearing.

Step 1

$$\text{Moment} = \frac{WL_e}{8} = 17 \times \frac{6.1}{8} = 12.96 \text{ kN m}$$

$$\text{Shear} = \frac{W}{2} = \frac{17}{2} = 8.5 \text{ kN}$$

Step 2

Normally, laminates are derived from 38 and 50 mm thicknesses and the code stipulates a maximum laminate depth of 44 mm. Finishing 38 mm would give 35 mm thickness and so we will consider both of these in the selected section.

By referring to Table 17 of the code, we obtain a breadth to depth ratio for restraint by purlins of $d/b = 4$; therefore, using this limitation we can arrive at a limiting depth with $b = 90$ mm $d = 90 \times 4 = 360$ mm maximum. With 44 mm laminates we obtain $360/44 > 8$ laminates; with 35 mm laminates we obtain $360/35 > 10$ laminates. Therefore

$$8 \times 44 = 352 \text{ mm}$$

or

$$10 \times 35 = 350 \text{ mm}$$

It will be shown later that the permissible bending stress and E values increase with the number of laminates used while beams over 300 mm in depth have a reduction factor applied to the bending stress.

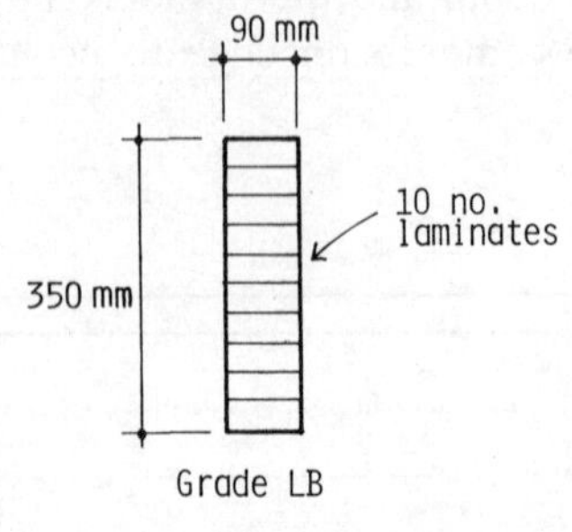

Fig. 9.7

As 10 No. laminates of 35 mm thickness give a smaller depth, this combination will be more favourable in design and will, therefore, be used. However, should this prove too expensive to manufacture the 44 mm laminates would supply an alternative. Three grades are given: LA, LB and LC but, of these, the most popular is LB because of its compatibility with available commercial grades. Therefore, summarizing Step 2 we have 35 mm thick laminates, a beam depth of 350 mm and a chosen grade of LB (Fig. 9.7).

Step 3

The code gives various modification factors which must be used in the design and which are related to a single grade used or a mixture of grades. As we are using a single grade, we will examine these values which affect our design.

K_1 = modification factor applied to the bending stress
K_2 = modification factor applied to the E value
K_4 = modification factor applied to the shear parallel to the grain
K_{16} = modification factor for the depth of beam if over 300 mm

From Table 5 of the code using LB grade and 10 No. laminates we have:

$$K_1 = 0.77 \qquad K_2 = 0.91 \qquad \text{and} \qquad K_4 = 0.90$$

$$K_{16} = 0.81 \times \frac{d^2 + 92\,300}{d^2 + 56\,800} \text{ (}d\text{ is expressed in millimetres)}$$

$$= 0.81 \times \frac{350^2 + 92\,300}{350^2 + 56\,800} = 0.97$$

In addition we may use an increase term factor of K_{12} because the beam is given as carrying roof loading. As such, medium term loading applies and the code permits an increase factor of $K_{12} = 1.25$. Our final consideration is the effect of end jointing of the laminate and its efficiency in bending.

For almost all glue laminated construction today, it will be found that staggered finger joints are used, but in addition CPP 112 gives efficiency ratings for plain scarf joints of slopes from 1 in 6 to 1 in 12. For this example, we will assume a 50 mm finger length whose profile complies with BS 5291:1976 and reference to Table 1 of this same code gives this particular joint an efficiency rating of 75%. To develop the full bending stress this efficiency rating must be at least equal to the product of $K_1 K_{16}$; if not, then the efficiency rating will replace this product.

$$K_1 \times K_{16} = 0.77 \times 0.97 = 0.747 < \text{efficiency rating of } 0.75$$

Hence 0.747 is the modification factor to be used.

Step 4

Now that we have the size and we know the various modification factors, we may determine the strength of the beam by finding its moment of resistance and shear capacity, following which we may check the applied deflection and bearing resistance.

Before we can find those values, we must arrive at the permissible design stress levels and in horizontal laminations this is done by taking the basic stresses for the timber species used and multiplying by the K value appropriate to the chosen grade. In this case, we will take for European Whitewood with SS grade selected for the E mean value.

Moment of resistance:

$$MR = f_{\text{ppar}} \times Z$$

where

$$f_{\text{ppar}} = f_{\text{bpar}} \times (K_1 \times K_{16}) \times K_{12}$$
$$= 14.5 \times 0.747 \times 1.25$$
$$= 13.54 \text{ N/mm}^2$$

and

$$Z = 9.0 \times \frac{35^2}{6} = 1837 \text{ cm}^3$$

giving

$$MR = 13.54 \times \frac{1837}{10^3} = 24.8 \text{ kN m} > 12.96 \text{ kN m applied.}$$

Shear capacity:

$$\bar{V} = v_p \times \tfrac{2}{3} bd$$

where

$$v_p = v_g \times K_4 \times K_{12}$$
$$= 1.52 \times 0.90 \times 1.25$$
$$= 1.71 \text{ N/mm}^2$$

giving

$$\bar{V} = 1.71 \times \tfrac{2}{3} \times 90 \times \frac{350}{1000}$$
$$= 35.9 \text{ kN} > 8.5 \text{ kN applied.}$$

Deflection: Because of the nature of the section and the method of manufacture it is possible to build in cambers to beams. However, in doing so, it may affect other features in the construction such as upper surface finishes and lining through with other members; therefore, pre-cambering should not be undertaken without some consideration. In this case, we will assume that pre-cambering is not permissible and so we must check the deflection for the total applied load as specified. Now

$$d = \frac{5WL^3}{384EI}$$

for a uniformly distributed load, where

$$E = E_{\text{mean}} \times K_2$$

$$= 10\ 000 \times 0.91 = 9100\ \text{N/mm}^2$$

and

$$I = \frac{bd^3}{12} = 9.0 \times \frac{35^3}{12} = 32\ 156\ \text{cm}^4$$

$$L_e = 6000 + 100\ (\text{c/c bearings}) = 6100\ \text{mm}$$

giving

$$d = \frac{5 \times 17\ 000 \times 6100^3}{384 \times 9100 \times 32\ 156 \times 10^4} = 17.2\ \text{mm}$$

$$\text{permissible } d_p = 6100 \times 0.003 = 18.3\ \text{mm} > 17.2\ \text{applied.}$$

The above deflection is caused by bending stresses but in addition to this there exists a further deflection caused by shear stresses. In timber design it is only a consideration where the beam can be said to be deep in relation to the span and it is normal practice to disregard this effect in solid rectangular sections or glue-laminated rectangular sections. The shear deflection at the centre span is given by the formula

$$d_v = \frac{FM_0}{AG} \tag{9.1}$$

where F is a form factor (1.2 for rectangular sections), M_0 is the bending moment at the centre span, A is the cross-sectional area, and G is the modulus of rigidity (usually $E/16$).

If we examine the shear deflection for this particular example we find that

$$d_v = \frac{1.2 \times 12.96 \times 10^4 \times 16}{9.0 \times 35 \times 9100} = 0.87\ \text{mm}$$

This accounts for approximately a 5% increase on the bending deflection making the total deflection 18.07 mm still within the permissible of 18.3 mm and, therefore, acceptable. However, this check does serve to illustrate how shear deflection can bring the total design deflection closer to the limit and, perhaps, in areas of sensitive finish, be the cause of unsightly cracking, should it be omitted. In ply-box beam construction, it can be critical and should always be checked (see later example Chapter 9.3).

Bearing: Our final check is the bearing stress over the supports where a modification factor of K_{13} may be adopted in accordance with Clause 3.12.1.2, Table 14 of the Code where it can be shown that the bearing occurs 75 mm or more from the end of the member. In this example the bearings occur at the ends of the beams and so K_{13} does not apply; also in glue-laminated construction, wane is not permitted, therefore, the basic stress is always used and not the grade stress.

$$\text{Bearing resistance} = Ab \times Cb_{\text{perp}} \times K_{12}$$

where

$$Ab = 10 \times 9 = 90\ \text{cm}^2$$

and

$$Cb_{\text{perp}} = 2.07\ \text{N/mm}^2$$

giving

$$\text{resistance} = \frac{90}{10} \times 2.07 = 18.63\ \text{kN} > 8.50\ \text{kN applied.}$$

9.3 Design of a ply web box beam

9.3.1 Introduction

Both ply web beams and ply web box beams (Figs 9.8 and 9.9, respectively) are commonly found in timber construction. They are generally used where heavy loads occur which cannot be carried by solid rectangular sections alone and where they occur in roof or floor systems, it is unusual for these beams to be spaced at centres closer than 1.2 m. The method of design most commonly used allows for the gross cross-sectional area of the ply where, for this purpose, the derived stress values have been reduced from the full value for parallel to bending stress to take account of the in-between veneers which are perpendicular to the bending stress.

Construction techniques include for overlapping butt end joints with splice plates in ply web beams, with suitable stiffening spaces, if so required by the design, while in the ply web box beam butt end joints are always backed by an internal vertical stud. Butt end joints should never be permitted to occur at the mid-span and so it is generally accepted that the centre portion will contain a full plywood sheet length placed asymmetric to the centre line of the span such that these joints may be staggered on each side of the section. Figure 9.10 illustrates this point.

If end shear reactions are high, it is normal to use additional pieces of plywood to each end, fixed to the outer faces in the case of the ply web or to either outer or inner faces in the case of box sections.

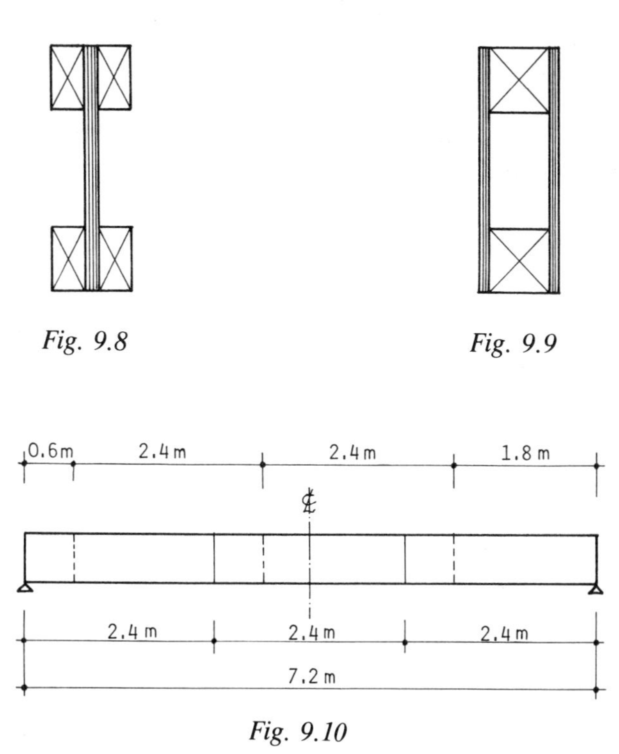

Fig. 9.8

Fig. 9.9

Fig. 9.10

Pre-cambering is often used to take account of the dead loads which means that the ply sheets will, for concealed beams, remain cut straight and only the chords are pre-cambered. If the beam is to be exposed, then it is normal to cut the plywood to the profile of the pre-camber.

Very rarely are plywood beams used where appearance is important because the manufacture is normally from unsanded plywood and squeezed glue lines are apparent. Therefore, finishing to an acceptable standard involves an additional cost. Assembly invariably involves the use of glue of WBP quality with pressure applied by nailing. Refer to Chapter 12, Sections 12.7.2 and 12.7.4.

9.3.2 Wood materials

The wood used to form the chords is largely of the species European Whitewood or Redwood with stress grades of numerical 50, SS or M75. Stiffeners may be of the same grade or of a lower quality down to GS standard. In regard to plywood, the three commonest types used are Douglas Fir-faced, Douglas Fir sheathing and Finnish birch-faced (Combi) and Tables 9.1 and 9.2 give the basic stresses. For both the Douglas firs, the face grain is parallel to the longest dimension while the Combi has its face grain perpendicular to the long side. Therefore, in the case of the former two, the grain is always parallel to the span whilst the latter is perpendicular to the span. Most designs adopt bending as the basis for the limiting stress values and so for the plywood the E value is that derived from bending in the flat plane, in the absence of any given values for bending in the vertical plane.

9.3.3 Method of design

There are several methods of approach to finding the moment capacity of a beam but for deflection one should always use the actual EI values for each of the two materials and sectional sizes used. The simplest and quickest way to determine the moment capacity is to assume a homogeneous section as in steel joist design and use the value of the bending stress for the grade of the chord material. This method tends to be a little liberal but in no sense could it be said to be too liberal, bearing in mind the degree of accuracy required. When checking a section, we need to examine the following applied values and compare with the permissible:

(a) bending stresses
(b) panel shear
(c) rolling shear
(d) end bearing
(e) deflection (bending plus shear)

To see exactly how this is done, let us work through a typical section for a ply web box beam.

Example 9.4

The proposed section for a ply web box beam is as indicated in Fig. 9.11. It is to be designed to carry a total roof load of 3.0 kN/m and is to span 7.2 m centre to centre of bearings. Verify the five salient design features as listed in Section 9.3.3 with the beams spaced at 2.4 m c/c.

Chords: European Whitewood to SS grade.
Webs: Douglas Fir-faced.

Step 1 Determine the moment of inertia and section modulus for a homogeneous section and hence check the moment capacity against the applied moment.
Step 2 Check rolling shear and panel shear stresses; determine support bearing depth required.
Step 3 Ascertain the total deflection and advise on pre-camber if any.

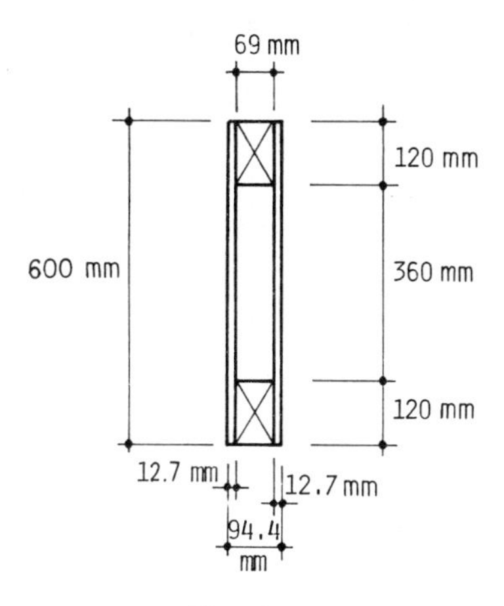

Fig. 9.11

Step 1

Assuming that a homogeneous section applies the moment of inertia can be related to the equation for two rectangles with the smaller internal void subtracted from the larger whole:

$$I = \frac{BD^3}{12} - \frac{bd^3}{12}$$

$$= 9.44 \times \frac{60^3}{12} - 6.9 \times \frac{36^3}{12} = 143\ 093\ \text{cm}^4$$

giving

$$Z = \frac{143\ 093 \times 2}{60} = 4770\ \text{cm}^3$$

Using the basic bending stress of 7.3 N/mm^2 for SS grade Whitewood and increasing by K_{12} for medium-term loading we have a permissible bending stress of:

$$f_{\text{ppar}} = 7.3 \times 1.25 = 9.125\ \text{N/mm}^2$$

Table 9.1 Dry grade stresses and moduli for nominal thicknesses in mm (inches) of Canadian Douglas Fir plywood

Type and direction of stress and modulus	Grade	*Grade stresses and moduli for the following nominal thicknesses (with the total number of plies in parenthesis)*													
		6.5 (1/4) (3) (N/mm²)	8 (5/16) (3) (N/mm²)	10 (3/8) (3) (N/mm²)	13 (1/2) (5) (N/mm²)	16 (5/8) (5) (N/mm²)	19 (3/4) (7) (N/mm²)	19 (3/4) (5) (N/mm²)	22.2 (7/8) (7) (N/mm²)	25.4 (1) (7) (N/mm²)	25.4 (1) (9) (N/mm²)	29 (9/8) (7) (N/mm²)	29 (9/8) (9) (N/mm²)	32 (5/4) (9) (N/mm²)	32 (5/4) (11) (N/mm²)
Extreme fibre in bending															
Face grain parallel to span (sanded grades)	Good 2 sides	11.2	—	9.93	7.03	5.93	6.48	—	5.58	5.38	7.72	5.52	7.10	5.65	6.62
	Good 1 side	10.3	—	9.10	6.41	5.45	5.93	—	5.10	4.96	7.03	5.03	6.48	5.17	6.07
	Solid 2 sides	9.58	—	8.48	6.00	5.10	5.58	—	4.83	6.62	6.62	4.76	6.07	4.83	5.65
	Solid 1 side														
Face grain parallel to span (unsanded grades)	Select sheathing	—	9.93	9.31	8.21	8.76	7.31	6.41	6.21	6.07	6.90	5.45	5.65	6.96	—
	Sheathing														
Face grain perpendicular to span	All grades sanded	2.4	—	4.55	6.07	6.69	5.72	—	6.41	6.41	4.21	6.14	4.69	5.93	5.24
	All grades unsanded	—	1.65	3.17	3.31	2.76	3.86	5.03	4.96	4.96	4.27	5.58	5.38	3.93	—
Tension															
Parallel to face grain (sanded grades)	Good 2 sides	8.41	—	6.21	5.79	5.38	5.24	—	5.38	4.76	7.24	5.52	6.62	5.58	5.58
	Good 1 side	7.72	—	5.72	5.45	5.03	4.90	—	5.03	4.41	6.76	5.17	6.21	5.24	5.24
	Solid 2 sides	7.17	—	5.31	5.45	5.03	4.90	—	5.03	4.41	6.76	5.17	6.21	5.24	5.24
	Solid 1 side														
Parallel to face grain (unsanded grades)	Select sheathing	—	8.14	6.48	6.55	7.38	6.83	6.00	4.96	5.65	5.72	4.96	4.76	6.76	—
	Sheathing														
Perpendicular to face grain	All grades sanded	3.93	—	5.45	4.90	5.24	5.38	—	5.24	5.79	3.72	5.17	4.21	5.03	5.03
Perpendicular to face grain	All grades unsanded	—	3.24	4.55	3.93	3.17	3.65	4.34	5.24	4.69	4.62	5.31	5.52	3.72	—
45° to face grain	Good 2 sides	1.86	1.86	1.86	1.86	1.86	1.86	1.86	1.86	1.86	1.86	1.86	1.86	1.86	1.86
	Good 1 side	1.79	1.79	1.79	1.79	1.79	1.79	1.79	1.79	1.79	1.79	1.79	1.79	1.79	1.79
	Solid 2 sides	1.72	1.72	1.72	1.72	1.72	1.72	1.72	1.72	1.72	1.72	1.72	1.72	1.72	1.72
	Solid 1 side	1.72	1.72	1.72	1.72	1.72	1.72	1.72	1.72	1.72	1.72	1.72	1.72	1.72	1.72
	Unsanded sheathing	1.72	1.72	1.72	1.72	1.72	1.72	1.72	1.72	1.72	1.72	1.72	1.72	1.72	1.72

Table 9.1 (continued)

Type and direction of stress and modulus	Grade	Grade stresses and moduli for the following nominal thicknesses (with the total number of plies in parenthesis)													
		6.5 (1/4) (3) (N/mm²)	8 (5/16) (3) (N/mm²)	10 (3/8) (3) (N/mm²)	13 (1/2) (5) (N/mm²)	16 (5/8) (5) (N/mm²)	19 (3/4) (7) (N/mm²)	19 (3/4) (5) (N/mm²)	22.2 (7/8) (7) (N/mm²)	25.4 (1) (7) (N/mm²)	25.4 (1) (9) (N/mm²)	29 (9/8) (7) (N/mm²)	29 (9/8) (9) (N/mm²)	32 (5/4) (9) (N/mm²)	32 (5/4) (11) (N/mm²)
Compression															
Parallel to face grain (sanded grades)	Good 2 sides	5.86	—	4.34	4.07	3.79	3.65	—	3.72	3.31	5.03	3.86	4.62	3.93	3.93
	Good 1 side	5.38	—	4.00	3.79	3.93	3.45	—	3.52	3.10	4.76	3.59	4.34	3.65	3.65
	Solid 2 sides	5.03	—	3.72	3.79	3.93	3.45	—	3.52	3.10	4.76	3.59	4.34	3.65	3.65
	Solid 1 side														
Parallel to face grain (unsanded grades)	Select sheathing	—	5.65	4.55	4.55	5.17	4.76	4.21	3.52	4.00	4.00	3.52	3.31	4.76	—
	Sheathing														
Perpendicular to face grain	All grades sanded	2.41	—	3.31	2.96	3.17	3.31	—	3.24	3.52	2.28	3.10	2.55	3.10	3.10
Perpendicular to face grain	All grades unsanded	—	2.00	2.76	2.41	1.93	2.21	1.69	3.24	2.83	2.83	3.24	3.38	2.28	—
45° to face grain	Good 2 sides	2.41	2.41	2.41	2.41	2.41	2.41	2.41	2.41	2.41	2.41	2.41	2.41	2.41	2.41
	Good 1 side	2.34	2.34	2.34	2.34	2.34	2.34	2.34	2.34	2.34	2.34	2.34	2.34	2.34	2.34
	Solid 2 sides	2.28	2.28	2.28	2.28	2.28	2.28	2.28	2.28	2.28	2.28	2.28	2.28	2.28	2.28
	Solid 1 side	2.28	2.28	2.28	2.28	2.28	2.28	2.28	2.28	2.28	2.28	2.28	2.28	2.28	2.28
	Unsanded sheathing	2.28	2.28	2.28	2.28	2.28	2.28	2.28	2.28	2.28	2.28	2.28	2.28	2.28	2.28
Bearing															
On face	All grades	2.62	2.62	2.62	2.62	2.62	2.62	2.62	2.62	2.62	2.62	2.62	2.62	2.62	2.62
Shear, rolling in plane of plies															
Parallel and perpendicular to face grain	All grades	0.345	0.345	0.345	0.345	0.345	0.345	0.345	0.345	0.345	0.345	0.345	0.345	0.345	0.345
45° to face grain	All grades	0.483	0.483	0.483	0.483	0.483	0.483	0.483	0.483	0.483	0.483	0.483	0.483	0.483	0.483

Table 9.1 (continued)

Type and direction of stress and modulus	*Grade*	*Grade stresses and moduli for the following nominal thicknesses (with the total number of plies in parenthesis)*													
		6.5 (1/4) (3) (N/mm²)	*8 (5/16) (3) (N/mm²)*	*10 (3/8) (3) (N/mm²)*	*13 (1/2) (3) (N/mm²)*	*16 (5/8) (5) (N/mm²)*	*19 (3/4) (7) (N/mm²)*	*19 (3/4) (5) (N/mm²)*	*22.2 (7/8) (7) (N/mm²)*	*25.4 (1) (7) (N/mm²)*	*25.4 (1) (9) (N/mm²)*	*29 (9/8) (7) (N/mm²)*	*29 (9/8) (9) (N/mm²)*	*32 (5/4) (9) (N/mm²)*	*32 (5/4) (11) (N/mm²)*
Panel shear															
Parallel and perpendicular to face grain	Good 2 sides	1.45	1.45	1.45	1.45	1.45	1.45	1.45	1.45	1.45	1.45	1.45	1.45	1.45	1.45
	Good 1 side	1.31	1.31	1.31	1.31	1.31	1.31	1.31	1.31	1.31	1.31	1.31	1.31	1.31	1.31
	Solid 2 sides	1.24	1.24	1.24	1.24	1.24	1.24	1.24	1.24	1.24	1.24	1.24	1.24	1.24	1.24
	Solid 1 side	1.24	1.24	1.24	1.24	1.24	1.24	1.24	1.24	1.24	1.24	1.24	1.24	1.24	1.24
	Unsanded sheathing	1.24	1.24	1.24	1.24	1.24	1.24	1.24	1.24	1.24	1.24	1.24	1.24	1.24	1.24
45° to face grain	Good 2 sides	2.90	2.90	2.90	2.90	2.90	2.90	2.90	2.90	2.90	2.90	2.90	2.90	2.90	2.90
	Good 1 side	2.69	2.69	2.69	2.69	2.69	2.69	2.69	2.69	2.69	2.69	2.69	2.69	2.69	2.69
	Solid 2 sides	2.48	2.48	2.48	2.48	2.48	2.48	2.48	2.48	2.48	2.48	2.48	2.48	2.48	2.48
	Solid 1 side	2.48	2.48	2.48	2.48	2.48	2.48	2.48	2.48	2.48	2.48	2.48	2.48	2.48	2.48
	Unsanded sheathing	2.48	2.48	2.48	2.48	2.48	2.48	2.48	2.48	2.48	2.48	2.48	2.48	2.48	2.48
Modulus of elasticity in bending															
Parallel to face grain	All grades sanded	11 300	—	10 100	7 310	6 270	6 830	—	5 930	5 790	7 930	5 860	7 380	6 000	6 900
	All grades unsanded	—	11 700	10 900	9 650	10 300	8 760	7 720	7 520	7 380	8 270	6 690	6 900	8 340	—
Perpendicular to face grain	All grades sanded	896	—	1 650	3 380	4 000	3 650	—	4 270	4 340	2 960	4 270	3 380	4 210	2 650
	All grades unsanded	—	621	1 100	1 860	1 520	2 480	3 100	3 240	3 310	2 760	3 790	3 650	2 760	—
Modulus of elasticity in tension and compression															
Parallel to face grain	All grades sanded	7 380	—	5 650	6 270	5 930	5 720	—	5 860	5 240	7 720	6 000	7 100	6 140	6 140
	All grades unsanded	—	8 270	6 760	7 450	8 340	7 790	6 960	5 860	6 550	6 620	5 860	5 580	7 650	—
Perpendicular to face grain	All grades sanded	3 310	—	4 410	4 000	4 270	4 340	—	4 270	4 690	3 100	4 210	3 520	4 140	4 140
	All grades unsanded	—	2 760	3 720	3 240	2 690	3 100	3 590	4 270	3 860	3 790	4 270	4 480	3 170	—
Modulus of rigidity															
Parallel and perpendicular to face grain	All grades	758	758	758	758	758	758	758	758	758	758	758	758	758	758
45° to face grain	All grades	2 480	2 480	2 480	2 480	2 480	2 480	2 480	2 480	2 480	2 480	2 480	2 480	2 480	2 480

Table 9.2 Dry grade stresses and moduli for Finnish European Birch plywood (finply-exterior)

Type and direction of stress and modulus	*Values of stress or modulus (N/mm²)*
Extreme fibre in bending	
Face grain parallel to span, 5 ply	15.9
Face grain parallel to span, 7 or more ply	13.8
Face grain perpendicular to span, 5 and 7 ply	8.41
Face grain perpendicular to span, 9 or more ply	10.8
Face grain 45° to span	8.55
Tension	
Parallel to face grain, 5 and 7 ply	14.5
Parallel to face grain, 9 or more ply	13.0
Perpendicular to face grain	9.58
45° to face grain	4.62
Compression	
Parallel to face grain, 5 and 7 ply	8.14
Parallel to face grain, 9 or more ply	7.17
Perpendicular to face grain	5.38
45° to face grain	4.90
Bearing	
On face	3.45
Shear, rolling in plane of plies	
Parallel and perpendicular to face grain	0.862
45° to face grain	0.862
Panel shear	
Parallel and perpendicular to face grain	3.10
45° to face grain	7.24
Modulus of elasticity in bending	
Face grain parallel to span, 5 ply	10 100
Face grain parallel to span, 7 or more ply	8 340
Face grain perpendicular to span, 5 and 7 ply	4 140
Face grain perpendicular to span, 9 or more ply	5 600
Face grain 45° to span	2 040
Modulus of elasticity in tension and compression	
Parallel to face grain	8 690
Perpendicular to face grain, 5 and 7 ply	7 100
Perpendicular to face grain, 9 or more ply	7 580
45° to face grain	2 180
Modulus of rigidity	
Parallel and perpendicular to face grain	800
45° to face grain	2 280

load sharing does not apply because the beam spacings exceed 610 mm. The moment capacity may now be found from

$$\bar{M} = f_{\text{ppar}} \times Z$$

$$= \frac{9.125 \times 4770 \times 10^3}{10^6} = 43.53 \text{ kN m}$$

applied moment

$$M = \frac{WL_e^2}{8} = 3.0 \times \frac{7.2^2}{8} = 19.44 \text{ kN m} < 45.53$$

which is satisfactory.

Step 2

The rolling shear stress (Section 1.3.2) is given by the formula

$$V_r = \frac{Va\bar{y}}{nd_c I}$$

Where

V = support maximum shear

$a\bar{y}$ = first moment of area of the chord about the neutral axis

n = number of contact faces between ply and chord

d_c = contact depth of chord

in this case

$$V = 3.0 \times \frac{7.2}{2} = 10.8 \text{ kN}$$

$$a\bar{y} = 6.9 \times 12.0 \times 24.0 = 1987 \text{ cm}^3$$

$$n = 2$$

$$d_c = 120 \text{ mm}$$

giving

$$v_r = \frac{10\,800 \times 198.7}{2 \times 120 \times 143\,093} = 0.062 \text{ N/mm}^2$$

The Code of Practice stipulates that in areas of concentrated rolling shear stress, as in the junction between plywood and chords in ply web beams, the permissible rolling shear must be reduced by 50%. This gives

$$\text{permissible} = 0.35 \text{ (Table 9.1)} \times 1.25 \times 0.5 = 0.218 \text{ N/mm}^2$$

which is satisfactory.

Panel shear stress (Section 1.3.1) is given by the formula

$$v_{pa} = \frac{VA\bar{y}}{tI}$$

where

$A\bar{y}$ = the first moment of area of the section above the neutral axis

t = the total thickness of the webs

in this case

$$A\bar{y} = 1.27 \times 2 \times 30 \times 15 + 1987\ (a\bar{y})$$
$$= 3130\ \text{cm}^3$$

giving

$$v_{pa} = \frac{10\ 800 \times 313}{2 \times 12.7 \times 143\ 093} = 0.93\ \text{N/mm}^2$$

permissible $= 1.31 \times 1.25 = 1.64\ \text{N/mm}^2$ satisfactory

The total end bearing area required will be governed by the compression stress in the whitewood perpendicular to the grain which is 1.55 N/mm^2 (Table 3.7). This may be increased by K_{12} to give a permissible stress of

$$C_{pper} = 1.55 \times 1.25 = 1.94\ \text{N/mm}^2$$

Hence the bearing area required

$$A_b = \frac{V}{C_{pper}}$$
$$= \frac{10\ 800}{1.94} = 5567\ \text{mm}^2$$

$$\text{minimum bearing depth} = \frac{5567}{94.4} = 59\ \text{mm}$$

It would be unusual to provide a seating less than 75 mm.

Step 3

The final stage in this design is to check the applied deflection against the permissible and in order to do this we must obtain the EI values for each of the materials used.

$E_{ply} = 9780\ \text{N/mm}^2$ $E_{chords} = E_N$ (see Example 9.2)

$E_N = E_2 = 6960\ \text{N/mm}^2$

$$I_{ply} = 2 \times 1.27 \times \frac{60^3}{12} = 457.2 \times 10^6\ \text{mm}^4$$

$$I_{chords} = \frac{6.9}{12}(60^3 - 36^3) = 973.7 \times 10^6\ mm^4$$

$$EI = 9780 \times 457.2 \times 10^6 + 6960 \times 973.7 \times 10^6$$
$$= 11\ 248 \times 10^9\ N/mm^2$$

$$\text{bending deflection } d = \frac{5WL_e^3}{384EI}$$

$$= \frac{5 \times (3000 \times 7.2) \times 7200^3}{384 \times 11\ 248 \times 10^9}$$

$$= 9.33\ mm$$

Now in addition to the deflection caused by bending we have that caused by the shear force.

In Example 9.3, equation (9.1), it is given that the shear deflection for a solid rectangular section at the centre span of simple beams is

$$d_v = \frac{FM_o}{AG} \text{ with } F = 1.2 \text{ for rectangular sections}$$

However, the shear deflection is affected by the position of the loads, the cross-sectional dimensions and the shear modulus of the webs. Roark (1956) makes the recommendation that F may be taken as unity provided that the cross-sectional area relates solely to the webs. This would give the modified equation of

$$d_v = \frac{M_o}{A_p G_p} \quad (9.2)$$

where

M_o = bending moment at the mid-span

A_p = cross-sectional area of the plywood

G_p = the modulus of rigidity of the plywood

Using (9.2) for the given example with

$M_o = 19.44$ kN m

$A_p = 1.27 \times 2 \times 60 = 152.4 \times 10^2\ mm^2$

$G_p = 807\ N/mm^2$

$$d_v = \frac{19.44 \times 10^4}{152.4 \times 807} = 1.58\ mm$$

and so the total deflection will be

$$d_t = d + d_v$$
$$= 9.33 + 1.58$$
$$= 10.91 \text{ mm}$$

This compares very favourably with the code's recommendation that the allowable deflection should be

$$d_p = 0.003 \times 7200 = 21.6 \text{ mm} > 10.91 \text{ mm}$$

In this particular case, it is unlikely that pre-cambering would be necessary unless aesthetics dictated in which case a pre-camber of around 10–15 mm would be more than satisfactory.

It is worth noting that for this example the shear deflection is approximately 14.5% of the total deflection which is not an uncommon figure in ply web designs and indicates the importance of considering shear deflection in these types of beams.

Sufficient information has been given in this chapter for the reader to be able to check the cross-section of a plyweb box beam and, of course, I-shape sections would follow the same procedure. There are, however, three other criteria which would require examination in order to complete the manufacturing information and these are

(a) determination of splice plate (butt-end joint) sizes
(b) design and spacing of web stiffeners
(c) design of end web stiffeners.

These designs follow accepted principles of theoretical analysis but will not be entered into here. Further reading is recommended by examining the COFI publication 'Fir plywood web beam design'.

9.4 Design of a flitched beam

9.4.1 Introduction

Very often in timber design there arises the situation where not only large spans and heavy loads predominate but also the available depth of the section is restricted in some way. When this does happen, the more conventional choice of sections invariably fails to produce a solution. Consequently, it is not unusual for the designer to investigate the introduction of a flitch beam.

A flitch beam consists of one or more pieces of flat plate steel (normally mild steel) sandwiched between two or more pieces of solid rectangular

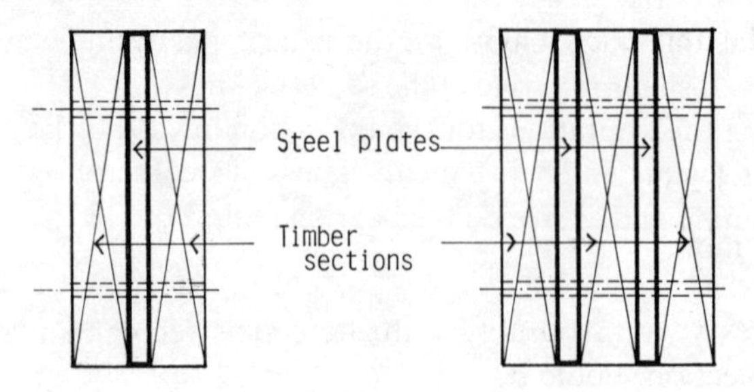

Fig. 9.12

timber (normally softwood) and the whole are bolted together at intervals in the beam's length (Fig. 9.12).

9.4.2 Design principles

The use of the modular ratio principle is not highly regarded in timber design whereas, for many years, it was used in the analysis of reinforced concrete sections where the materials have differing E values. As in reinforcing concrete, the object is to obtain a section which is stronger than the basic material, in this case the timber, at an economical level.

In order to achieve this objective, the reinforcing agent must have a significantly larger E value than the reinforced material. In the case of flitch beams, this is easily obtained in that the modular ratio is invariable between 1:20 and 1:30. The principle of the modular ratio design method is to check the strength of the section against the permissible stresses of the weakest material using the modular ratio as an increase factor when dealing with the sectional properties of the steel plate. This is best exemplified in the detailed example which follows.

Example 9.5

A flitch beam is to be inserted within the floor depth of a domestic house. It carries an external wall and roof loading from the upper storey section and the total load may be taken as 100 kN. The beam will span an effective distance of 4.8 m and will be formed from two 50×194 floor joists with a steel plate sandwiched between (Fig. 9.13). Determine the minimum thickness of steel plate required and check the deflections.

$E_{steel} = 210\ 000\ \text{N/mm}^2$, E_{timber} (Whitewood) $= E_N$ (SS grade), $E_N = E_2 = 6960\ \text{N/mm}^2$.

Step 1 Determine the equation for the moment capacity of the section and compare with the applied moment to arrive at the steel thickness.

Step 2 Determine the equation for the shear capacity and compare with the applied shear to arrive at the steel thickness.

Step 3 Using the limitations for the deflection in CP 112:Part 2, determine the EI value and hence the thickness of steel required.

Step 4 Examine and advise on the bearing required.

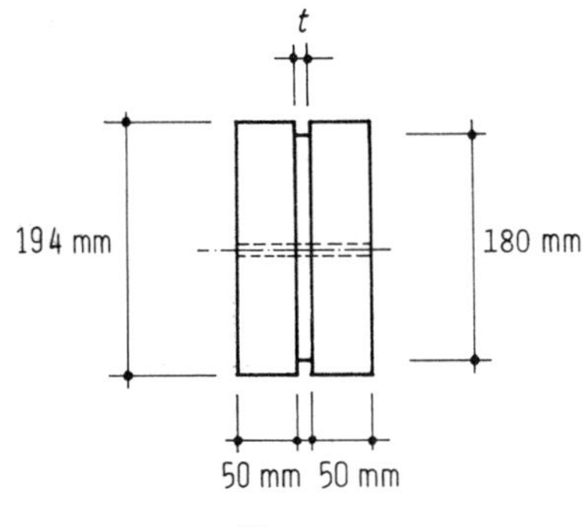

Fig. 9.13

Step 1

The modular ratio is

$$\frac{E_s}{E_t} = \frac{210\,000}{6960} = 30$$

Now the moment capacity $\bar{M} = fZ$ which for the total section may be written as

$$\bar{M} = f_{\text{ppar}} Z_t + 30 f_{\text{ppar}} Z_s \tag{9.3}$$

where

$$\bar{M} = M = \frac{WL}{8} = 100 \times \frac{4.8}{8} = 60.0\ \text{kN m}$$

$$Z_t = 2 \times 5.0 \times \frac{19.4^2}{6} = 627\ \text{cm}^3$$

$$Z_s = t \times \frac{18.0^2}{6} = 54t\ \text{cm}^3$$

$$f_{\text{ppar}} = f_{\text{gpar}} \times K_{12}$$

$$= 7.3 \times 1.25 = 9.125\ \text{N/mm}^2$$

Now substituting into (9.3) we have

$$60.0 \times 10^6 = 9.125 \times 627 \times 10^3 + 30 \times 9.125 \times 54t \times 10^3$$

$$t = \frac{60\,000 - (9.125 \times 627)}{30 \times 9.125 \times 54}$$

$$= 3.7 \text{ mm to satisfy bending}$$

Step 2

The equation for the shear capacity is

$$\bar{V} = v_{ppar} A_t \tfrac{2}{3} + 30 v_{ppar} A_s \qquad (9.4)$$

where

$$A_t = 5.0 \times 19.4 \times 2 = 194 \text{ cm}^2$$

$$A_s = 18t \text{ cm}^2$$

$$v_{ppar} = v_{gpar} \times K_{12}$$

$$= 0.86 \times 1.25 = 1.075 \text{ N/mm}^2$$

$$\bar{V} = V = 100 \times 0.5 = 50 \text{ kN}$$

Substituting into (9.4)

$$50 \times 10^3 = 1.075 \times 194 \times 10^2 \times \tfrac{2}{3} + 30 \times 1.075 \times 18t \times 10^2$$

$$t = \frac{50\,000 - 13\,903}{58\,050}$$

$$= 0.622 \text{ mm to satisfy shear}$$

Step 3

The equation for deflection is that for a uniformly distributed load

$$d = \frac{5WL^3}{384EI} \qquad (9.5)$$

with

$$L = 4800 \text{ mm}$$

$$d = d_p = 0.003 \times 4800 = 14.4 \text{ mm}$$

$$W = 100 \text{ kN}$$

Substituting into (9.5) and transposing

$$EI = \frac{5 \times 100 \times 10^3 \times 4800^3}{384 \times 14.4}$$

$$= 1.0 \times 10^{13}\ \text{N mm}$$

Now

$$EI = E_N I_t + E_s I_s \tag{9.6}$$

where

$$I_t = 2 \times 5.0 \times \frac{19.4^3}{12} = 6084\ \text{cm}^4$$

$$I_s = t \times \frac{18^3}{12} = 486t\ \text{cm}^4$$

Substituting into (9.6)

$$1.0 \times 10^9 = 6960 \times 6084 + 210\ 000 + 486t$$

$$t = \frac{1000 - 42.3}{102}$$

$$= 9.4\ \text{mm to satisfy deflection}$$

By comparing these three results it can be seen that the limitation is the thickness required to satisfy the deflection and so a minimum 10 mm thick plate should be provided.

Step 4

As the permissible bearing stress is influenced only by wane (see Section 3.2.5) it is possible to use the basic stress in compression perpendicular to the grain provided that the specification precludes the defect of wane over the supports. Assuming this to be the case then

$$C_{\text{ppar}} = C_{\text{bpar}}\ (\text{Table 3.2}) \times K_{12}$$

$$= 2.07 \times 1.25 = 2.58\ \text{N/mm}^2$$

Hence the required bearing area

$$A_b = \frac{500}{2.58} = 194\ \text{cm}^2$$

giving bearing depth

$$D = \frac{194}{10} = 19.4 \text{ cm}$$

A minimum seating at each end of the beam of 200 mm would satisfy the design and the ends of the timber should be trimmed such that the bearing surface is flush with the underside of the steel. In effect, this would reduce the total shear capacity of the timber because of the reducing effect of the notch cut (introducing also K_{14} of CP 112); however, as the steel thickness provided is well in excess of that required (10 mm > 0·622 mm) no further check is necessary.

To summarize, the final specification for this beam could be written as follows:

PROVIDE: (1) 10 mm thick × 180 mm deep mild steel plate pre-drilled on centre line with 4 No. 12.5 mm diameter holes positioned 600 mm in from each end and at 1.2 m centre spacings. Finished with two coats red oxide or similar.
(2) 2 No. 50 × 194 European Whitewood sections to minimum SS grade bolted one each side of steel plate.
(3) 4 No. 12 mm ϕ × 127 mm galvanized hex bolts with 50 mm ϕ × 3 mm washers beneath head and nut.

Note that the bolts should be checked for shear occurring in the beam to ensure composite action between the steel plate and timber members.

9.5 Roof purlin beams

9.5.1 Types

Generally speaking there are two types of purlin beams: those which are installed normal to the slope of the roof and those installed in a vertical plane. In the latter the justification follows those principles outlined in Example 9.1 except that because purlins are spaced at larger centres than 600 mm, load sharing would not be permitted and the E value would reduce, in the case of solid sections, to E_{min}.

For the first type two directions of load must be considered: one normal to the roof slope and the other caused by a component of the normal load acting down the slope and across the top surface of the purlin as indicated in Fig. 9.14.

The load W_1 is considered in relation to axis X–X whilst W_3 is taken to act upon the weaker axis Y–Y. It can be seen that there are two points of maximum bending stress occurring, one at each extreme corner of the section along the vertical plane Z–Z.

Therefore the maximum bending stress is a combination of axes X–X and Y–Y and equates to:

$$f_{apar} = \frac{M_{XX}}{Z_{XX}} + \frac{M_{YY}}{Z_{YY}} \tag{9.7}$$

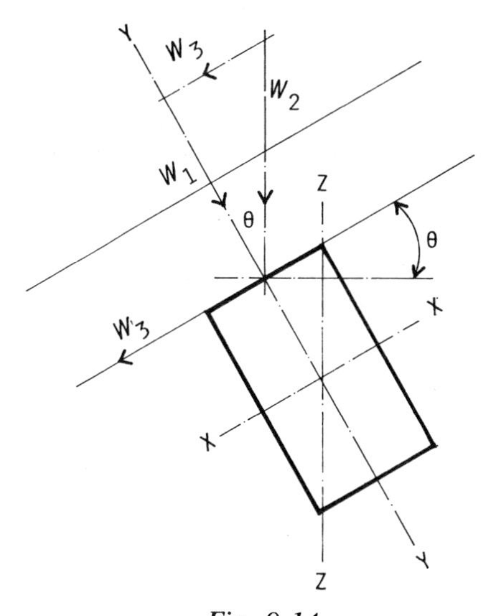

Fig. 9.14

For deflection the maximum displacement occurs through the axis Z–Z and its value may be simplified to:

$$d_{ZZ} = \sqrt{(d_{XX}^2 + d_{YY}^2)} \tag{9.8}$$

The successful design of this type of beam is largely a question of trial and error.

9.5.2 Choice of type

Wherever possible a purlin beam should be placed in the vertical plane and if necessary the supported construction should be notch seated over. This very much simplifies the design and gives an economical section. However, where purlins are placed normal to the roof slope, careful attention must be given to the end support where possible sliding and rotation can occur. Where purlins span across rafter backs, end blocking pieces, placed on the lower slope side, are a common feature. This blocking piece must of course be able to withstand the total shear load of W_3. Another method is to build the end into brick or block supporting walls which achieves the same effect as the blocking piece, but it should be remembered that there will be a tendency for shrinkage to occur and also treatment of the ends may be necessary to avoid rot. In conclusion, the choice of purlins normal to the roof slope will always be more expensive than the vertical purlin.

Example 9.6

A roof purlin spanning 3.6 m is positioned normal to a roof slope of 25° and the total load that it carries on plan is 10.0 kN. Determine a suitable size assuming that the top surface is fully restrained and the ends cannot rotate or slide.

Step 1 Determine the total loads W_1 and W_3 (see Fig. 9.14).
Step 2 Assume a size and determine f_{apar} from equation (9.7).
Step 3 Check for the maximum deflection using equation (9.8).

Step 1

Referring to Fig. 9.14 it can be seen that $\theta = 25°$ and $W_2 = 10.0$ kN from which

$$W_1 = W_2 \cos\theta$$
$$= 10.0 \times \cos 25°$$
$$= 9.06 \text{ kN}$$
$$W_3 = W_2 \sin\theta$$
$$= 10.0 \times \sin 25°$$
$$= 4.23 \text{ kN}$$

Step 2

Assuming a size of 150 mm × 250 mm sawn section in Whitewood we must first determine the properties and bending stress about each axis.

Axis X – X:

$$M_{XX} = W_1 \frac{Le}{8}$$

where $Le = 3.6$ m

$$= 9.06 \times \frac{3.6}{8} = 4.08 \text{ kN m}$$

$$Z_{XX} = 15.0 \times \frac{25.0^2}{6} = 1562 \text{ cm}^3$$

$$I_{XX} = 1562 \times \frac{25.0}{2} = 19\ 525 \text{ cm}^4$$

$$f_{apar} = \frac{4080}{1562} = 2.62 \text{ N/mm}^2$$

Axis Y–Y:

$$M_{YY} = W_3 \frac{Le}{8}$$

$$= 4.23 \times \frac{3.6}{8} = 1.90 \text{ kN m}$$

$$Z_{YY} = 25.0 \times \frac{15.0^2}{6} = 937.5 \text{ cm}^3$$

$$I_{YY} = 937.5 \times \frac{15.0}{2} = 7031 \text{ cm}^4$$

$$f_{apar} = \frac{1900}{937.5} = 2.03 \text{ N/mm}^2$$

$$\text{Using } \sum f_{apar} = \frac{M_{XX}}{Z_{XX}} + \frac{M_{YY}}{Z_{YY}} \qquad (9.7)$$

$$\sum f_{apar} = 2.62 + 2.03$$

$$= 4.65 \text{ N/mm}^2$$

$$f_{ppar} = 7.3 \times 1.25 = 9.125 \text{ N/mm}^2 > 4.65$$

Step 3

The maximum deflection in each direction must first be determined before equation (9.8) can be applied.

$$d_{XX} = \frac{5W_1 Le^3}{384EI_{XX}}$$

$$= \frac{5 \times 9060 \times 3600^3}{384 \times 5700 \times 19\ 525 \times 10^4} = 4.94$$

$$d_{YY} = \frac{5W_3 Le^3}{384EI_{YY}}$$

$$= \frac{5 \times 4230 \times 3600^3}{384 \times 5700 \times 7031 \times 10^4} = 6.41$$

Using $d_{ZZ} = \sqrt{(d_{XX}^2 + d_{YY}^2)}$ (9.8)

$$d_{ZZ} = \sqrt{(4.94^2 + 6.41^2)}$$

$$= 8.09 \text{ mm}$$

$$d_p = 0.003 \times 3600 = 10.8 \text{ mm} > 8.09$$

Therefore it is shown that a 150 mm × 250 mm sawn section in European Whitewood or Redwood to SS grade will satisfactorily carry the load of 10.0 kN.

9.6 Commentary on lateral stability for deep beams

9.6.1 Description

The description of a deep beam is largely empirical and CP 112 covers solid and glulam beams by calling for differing amounts of lateral restraint measured against specified breadth-to-depth ratios. This same code also gives guidance in the case of ply-box and I beam sections by using the ratio of I_x to I_y and comparing against similar methods of restraint as those for solid sections. In both cases, if the rules are followed, it is assumed that the full bending stresses can be developed. Exceeding those rules could lead to smaller permissible bending stresses.

What is clear to all timber designers is that if a beam becomes deep in relation to its width, i.e. visually disturbing (in an engineering sense), lateral restraint becomes a prime consideration and must be examined at both upper and lower levels of the beam's depth.

9.6.2 Design

Buckling in deep beams is influenced by several items:

(a) The shape of the beam's cross-section
(b) The type and position of loading applied
(c) The position of (b) in regard to the beam's neutral axis
(d) The amount of lateral restraint provided.

As previously stated, CP 112 uses for solid sections depth-to-breadth ratios D/b to define limitations for degrees of lateral stability, however it is known that the true relationship is LD/b^2 and consequently CP 112 can be misleading where lateral stability becomes a feature of design.

Where beams have a small stiffness in the direction of the weaker axis it is possible for failure to occur at smaller bending stresses than the design may indicate. Loading may reach a critical stage without undue signs of stress occurring and yet the beam is in a state of instability which can be manifested by immediate buckling at the slightest load increase.

The solution is to determine either the critical bending moment or critical bending stress at which failure will occur and employing a suitable factor of safety (normally 2.25), determine a permissible buckling moment or bending stress which must be equal to or greater than the bending moment or bending stress about the major axis produced by the design loading.

The length and detail of the particular subject is beyond the scope of this book. However, there is much in the way of publications, and formulae for buckling moments are given in Ozelbon and Baird's *Timber Designer's Manual* with tables for various coefficients related to common loading conditions. Similarly the critical bending stress approach is discussed in a paper prepared for The International Council for Building Research Studies and Documentation by H. J. Larson and entitled 'The design of timber beams' (October, 1975). Both publications are recommended for further reading on this particular subject.

10

Examples of element design – Part II

10.1 Axially loaded solid section posts

For this part, we will assume that a prop column is defined as a vertical piece of timber pinned or fixed at each end, carrying a load applied at the upper end and free from obstructions on all four sides as indicated in Fig. 10.1.

We will go on to discuss the following features and how they affect the design of a prop column:

End restraints
Lateral restraint
Effective length
Slenderness ratio
Bearing

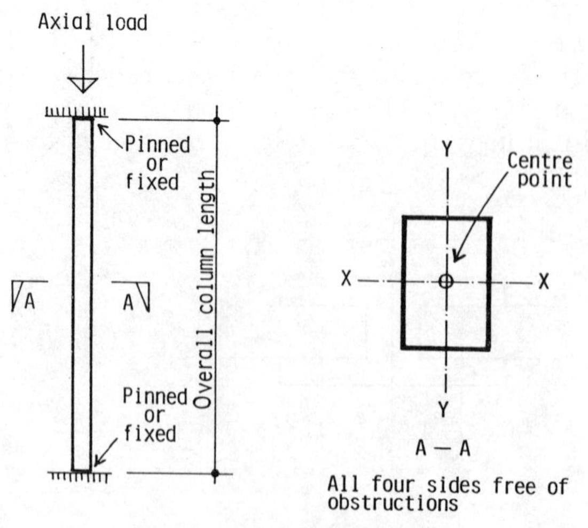

Fig. 10.1

The title of the subject refers to an 'axial' load which may be defined as a load which will act on the centre point (centre of gravity) of the cross-section of the timber (see Section A–A above). An eccentrically applied load is one which is applied at a given distance from the centre point of the section and thus will induce bending about one or both of the axes X–X and Y–Y. If the distance is called e then the moment is as indicated in the sketches (Fig. 10.2). The inducement of such eccentric loading in timber columns should always be avoided if only because of the possible long-term distortion which could follow.

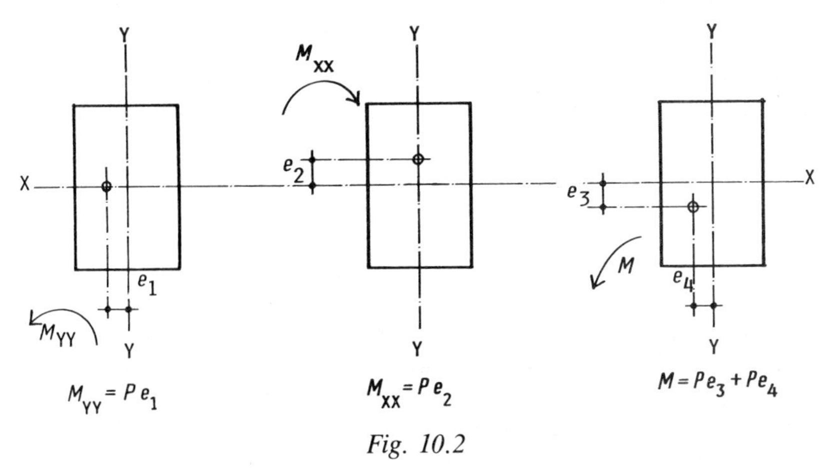

Fig. 10.2

An example of this form of eccentricity is a column head providing a shared bearing for two or more beams which carry different magnitudes of total load giving differing end support reactions (Fig. 10.3). The net result being an eccentric load acting about one or both of the principal axes of bending. In this case, it would be better to allow the beam, if at all possible, to span unbroken across the head of the column. Alternatively, provide two separate posts side by side such that they are each supporting only one of the loads. Having

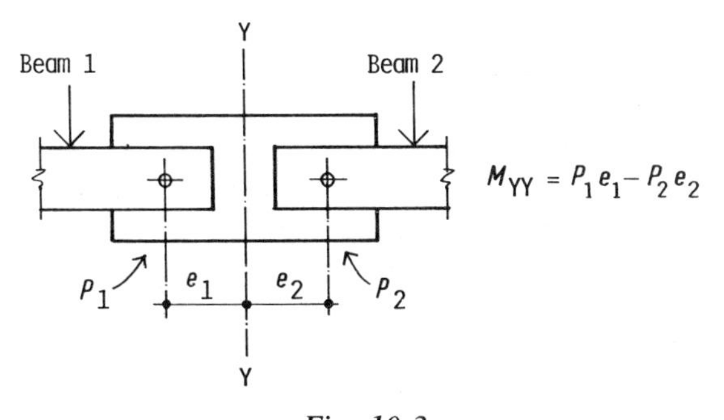

Fig. 10.3

established the importance of obtaining an axial load, we may now continue and discuss the relevance of each listed feature which, together, form the whole solution of the design.

10.1.1 End restraints

End restraints fall into two categories:

(a) positional – held in line but free to rotate, i.e. pinned (Fig. 10.4).
(b) positional and directional – held in line and against rotation, i.e. fixed (Fig. 10.5).

The correct interpretation of the end restraints is an important feature of the design because, as will be shown, it governs the chosen effective length, the

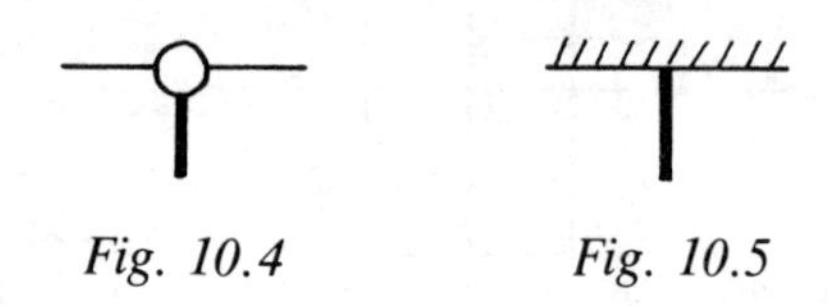

Fig. 10.4 *Fig. 10.5*

slenderness ratio and, consequently, the ultimate load which may be carried by the column. Determination of end restraints is a matter of individual interpretation primarily born of much variation in experience. The assumptions of one designer may not necessarily satisfy another and so, having chosen, the designer should be prepared to defend his choice with sound reasoning. In the event of doubt, then always choose pinned at both ends and take the full height of the column as its effective length. As a precautionary measure, this procedure cannot be faulted. In this book, the effects of sway are ignored and it is assumed that all ends remain in line with no end connection moving relative to the other.

10.1.2 Lateral restraint

Lateral restraint is a means by which the effective length between the chosen end restraints may be reduced and so increase the load-carrying capabilities of the column. For example, a column subjected to bending about its strongest axis will be governed by the tendency for the section to want to move in the direction of its weakest axis (known as buckling). This movement may be much reduced by laterally restraining the column in its weak axis direction. A good example of this is to be found in timber roof truss designs where binders are sometimes provided to laterally restrain members in compression against buckling (Fig. 10.6).

Such members are in effect columns subject to axial load and CP 112:Part 3:1973, the relevant governing code of practice, provides for assumed end

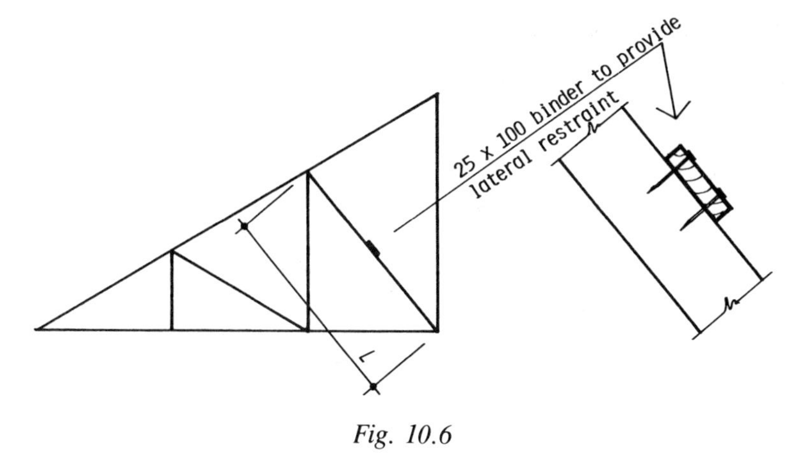

Fig. 10.6

restraints by allowing an effective length of $0.9L$. As a binder has been introduced, the final effective length for the compression member in this illustration would be $0.9L \times 1/2$ because the binder divides the effective length in half. However, it should be noted that these binders must also be restrained by some part of the total structure.

10.1.3 Effective length

The effective length of a column is that length which, after determination of the degree of the end restraints, dictates the load-carrying capacity of the section. CP 112:Part 2:1971, Clause 3.14.1, Table 18, gives the coefficients by which the actual length (L) of the member should be multiplied to arrive at the effective length.

For the work covered in this book and, indeed, for the majority of design cases in practice, the first three coefficients given in Table 18 will suffice and the diagrams in Fig. 10.7 indicate how they may be interpreted against the actual problem.

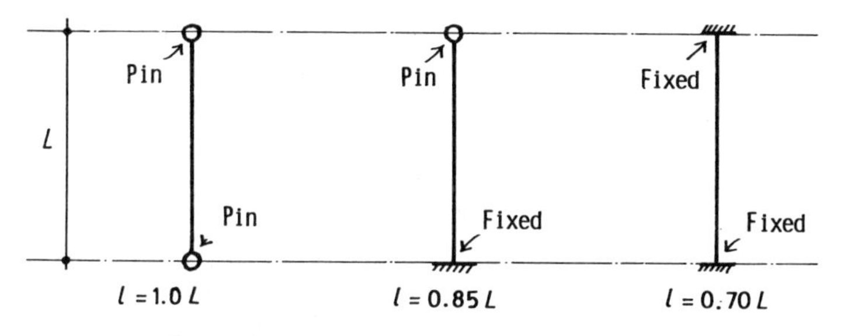

Fig. 10.7

10.1.4 Slenderness ratio

This is the ratio which is obtained by dividing the effective length by the least radius of gyration (r). In many cases, a column will have different effective lengths in relation to the major and minor axes and so both ratios will have to be calculated so that the critical one is properly used. For example in the column in Fig. 10.8 it can be seen that the two slenderness ratios are:

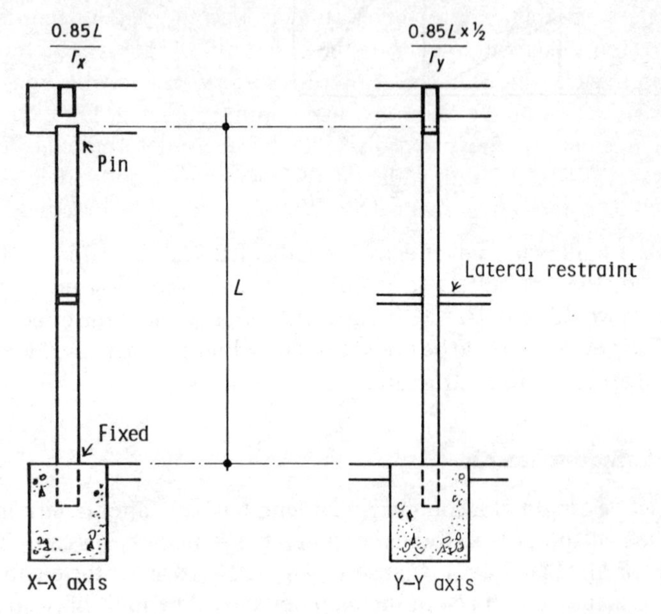

Fig. 10.8

For the columns which we will consider, CP 112:Part 2:1971, Clause 3.14.2 stipulates a maximum permissible slenderness ratio of 180 for any member carrying loads induced by dead and imposed loads. In the case of a column subjected to an axial load created solely by wind action, this ratio may increase to 250.

10.1.5 Bearing

The permissible bearing stress at the top and bottom of a column is the permissible grade stress parallel to the grain with $L/r = 0$ (C_{ppar}). A load sharing increase factor may be used as appropriate. However, when the column bears on a sole plate or the load is applied through a header binder, then the permissible stress is the grade stress perpendicular to the grain (C_{pper}). Term and load bearing increase factors apply. Also, where wane is excluded, the full cross-section width may be used in assessing the stress.

The load-carrying capability of a timber column is based upon the slenderness ratio which, as described beforehand, is a function of the cross-section and its effective length. But the slenderness ratio cannot be determined until the cross-section is known and so the actual design process is one of trial and error. It is possible to derive safe load charts for given cross-sections and various selected lengths but the purpose of this exercise is to give substance to the topic, thereby allowing the designer to obtain his own particular column size. Understandably, reference charts may become part of the designer's procedure but this should never be the case until the subject matter is fully understood. More often than not, there is some criterion which enables the column size to be found without too much initial guessing. For example, the stud size in a load-bearing partition is inevitably pre-determined by either an architecturally selected wall width or the most easily obtainable supplier's section. Most commonly, architectural restraints give the immediate guide to the solution. The two examples which follow best serve to show how such simple columns are designed.

In these examples the values of K_{18} are taken from Table 15 of CP 112, Part 2:1971 where K_{18} is a factor which allows for the effect of duration of loading on compression members of softwood depending upon the slenderness ratio. Similarly K_{19} allows for the same effect when using hardwoods.

Example 10.1

An existing rough sawn 100 mm square section timber prop column is 2.70 m high overall. It may be assumed that it carries an axial load and is theoretically pinned at each end. The timber is Whitewood and was previously visually stress graded to Appendix A of CP 112:Part 2:1971, to give 50 grade, and dry stresses are permissible. It is proposed to apply additional load to the building which will increase the existing axial load from 20.0 kN to 25 kN. The column is in an isolated position, carries no restraint in any direction and supports part of an office floor construction. Determine if the column is suitable.

Step 1

Draw a line diagram of the column, Fig. 10.9, and determine the properties of the cross-section. As the section is square the $I_{XX} = I_{YY}$ and so the slenderness ratio is the same for both axes.

$$r = \sqrt{\frac{I}{A}} \qquad I = 10 \times \frac{10^3}{12} = 833.3\ \text{cm}^4 \qquad A = 100\ \text{cm}^2$$

Hence

$$r = \sqrt{\frac{833.3}{100}} = 2.88\ \text{cm}$$

Step 2

Determine the slenderness ratio. With pinned supports at each end and no lateral restraint,

$$l = L = 2.70 \text{ m}$$

$$\text{and so } \frac{l}{r} = \frac{270}{2.88} = \text{say } 94$$

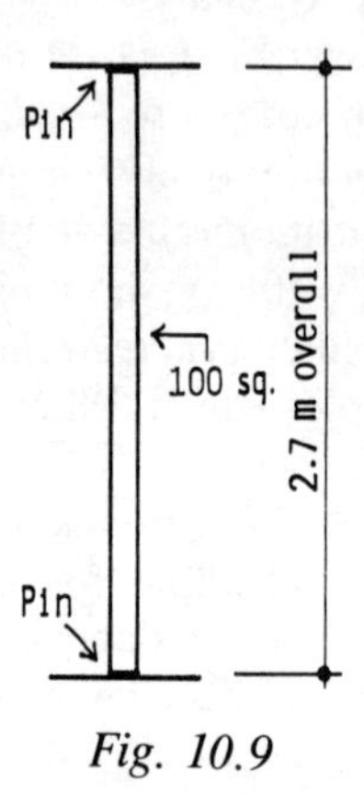

Fig. 10.9

Step 3

Knowing l/r obtain K_{18} and so calculate the maximum permissible design stress. As the column supports office loading, it must be taken to be of long-term duration and so from Table 15 of CP 112, Part 2, we have $K_{18} = 0.578$.

The column is isolated, therefore load sharing does not apply, thus the permissible design stress is the grade stress parallel to the grain given in Table 3.2 multiplied by the reduction factor of K_{18}, i.e.

$$C_{\text{gpar}} \text{ (50 grade)} = 4.8 \text{ N/mm}^2$$

$$C_{\text{ppar}} = 4.8 \times 0.578 = 2.77 \text{ N/mm}^2$$

Step 4

Having now determined the maximum permissible stress, it is a simple matter to find the maximum load which the column can carry, this being AC_{ppar}

$$\text{maximum permissible load} = \frac{100 \times 10^2 \times 2.77}{10^3} = 27.7 \text{ kN}$$

The proposed load is 25.0 kN and so the existing column section is satisfactory.

Example 10.2

Each post in an open stud wall shown in Fig. 10.10 has to resist a concentric axial load of 15 kN from a roof structure. SS grade Redwood is to be used and the timber is to be treated because of its exposed condition. Find the nearest available sawn section given that the width of the timber must not exceed 100 mm and check the bearing stress at the header.

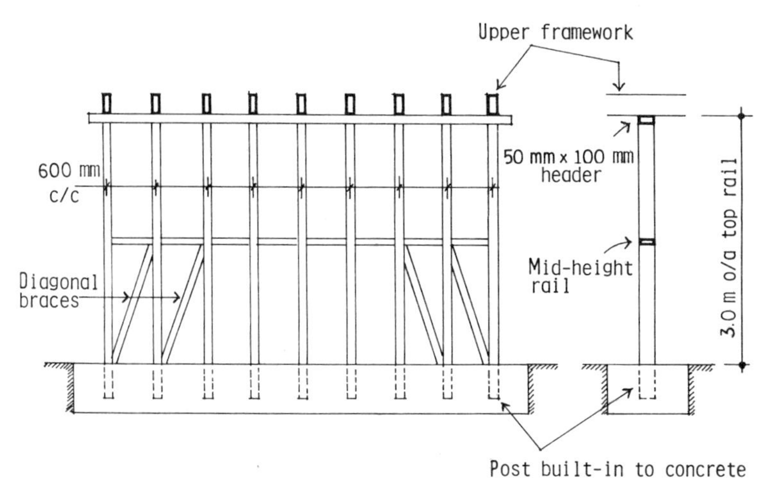

Fig. 10.10

Step 1

Draw a line diagram of the loaded post and determine its effective length (Fig. 10.11). The sketch of the fence indicates that the head of the post is in a pinned condition (restrained in position but not in direction) while the foot is fixed (restrained in both position and direction). Lateral restraint may be assumed at right-angles to the 100 mm face because the mid-height rail is adequately braced at each end. From CP 112:Part 2, Table 18,

$$\text{Effective } l = 0.85L$$

$$= 0.85 \times (3000 - 50)$$

$$= 2508 \text{ mm}$$

However, this effective length will only apply in the direction across the 100 mm face which we will call axis X–X. For the Y–Y axis, with a mid-height lateral restraint the effective length will be half that of l_x, i.e.

$$l_y = \frac{2508}{2} = 1254 \text{ mm}$$

Step 2

Assume a thickness and determine the slenderness ratios for both axes, thus find the limiting applied design stress and compare with the permissible.

Try 38 × 100: From standard property tables

$$r_x = 28.9 \text{ mm} \qquad r_y = 11.0 \text{ mm}.$$

$$\frac{l_x}{r_x} = \frac{2508}{28.9} = 87 \qquad \frac{l_y}{r_y} = \frac{1254}{11.0} = 114$$

Hence l_y/r_y governs.

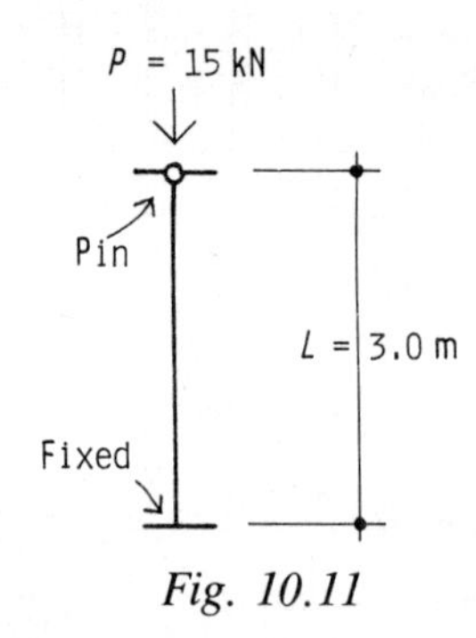

Fig. 10.11

With a roof structure medium term loading applies giving $K_{18} = 0.468$ and using green stresses because of the exposed nature of the work we have from Table 3.6

$$C_{\text{gpar}} \text{ (SS grade)} = 5.4 \text{ N/mm}^2.$$

Load sharing may be taken because the centres of the posts are at a spacing which does not exceed 610 mm c/c. The final permissible design stress will therefore be

$$C_{\text{ppar}} = 5.4 \times 1.1 \times 0.468 = 2.78 \text{ N/mm}^2$$

Finally, the applied stress

$$C_{\text{apar}} = \frac{P}{A} = \frac{150}{3.8 \times 10} = 3.95 \text{ N/mm}^2 > 2.78$$

It follows that this first trial section is not acceptable and another will need to be examined.

Try 50 × 100 sawn:

$$r_x = 28.9 \text{ mm (as before)}$$

but now

$$r_y = 14.4 \text{ mm}$$

$$\frac{l_y}{r_y} = \frac{1254}{14.4} = 87$$

which is the same as $\frac{l_x}{r_x}$

$$K_{18} = 0.713 \text{ giving}$$

$$C_{\text{ppar}} = 5.4 \times 1.1 \times 0.713 = 4.23 \text{ N/mm}^2$$

$$C_{\text{apar}} = \frac{150}{5 \times 10} = 3.0 \text{ N/mm}^2 < 4.23 \qquad \text{Satisfactory}$$

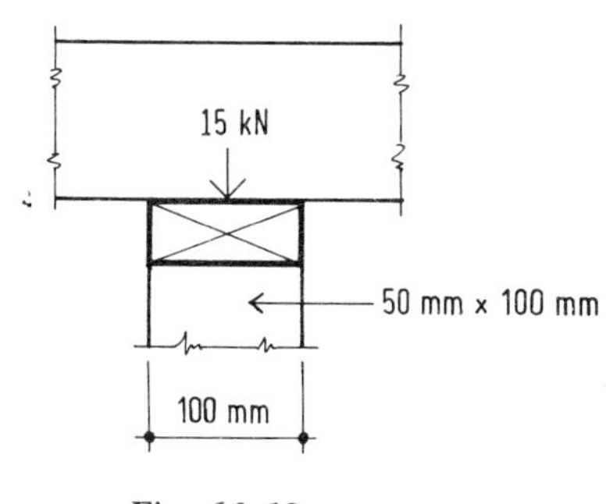

Fig. 10.12

Check the bearing stress at the junction with the header assuming that wane is not permitted (Fig. 10.12).

Permissible bearing stress is the stress perpendicular to the grain in the header, i.e.

$$C_{\text{aperp}} = \frac{15\,000}{50 \times 100} = 3.0 \text{ N/mm}^2$$

$$C_{\text{gperp}} = 1.14 \text{ N/mm}^2$$

Load sharing and medium-term K_{18} increase factors apply. In addition K_{13} for the length and position of the bearing may also be used and, in this case, as we are checking the limiting bearing stress related to the header, the length of the bearing is in effect the timber thickness of 50 mm.

Hence from Table 13, of CP 112, K_{13} is found to be 1.20 giving

$$C_{\text{pperp}} = 1.14 \times 1.1 \times 1.25 \times 1.20 = 1.88 \text{ N/mm}^2 < 3.0 \text{ N/mm}^2$$

This permissible stress will also apply to the end studs in the wall because the header overlaps by 75 mm meeting the requirements of Clause 3.12.1.2. Without this oversailing dimension the end stud stress would govern the design because K_{13} would not apply.

Even so, this check indicates that the bearing stress is too high and so it would appear that the easiest solution would be to introduce a hardwood header. Therefore, we will try Keruing to a 50 grade green stress.

From Table 3.1

$$C_{\text{pperp}} = 2.34 \times 1.1 \times 1.25 \times 1.2 = 3.86\ \text{N/mm}^2 > 3.0\ \text{N/mm}^2$$

PROVIDE: 50 × 100 SS grade Redwood posts at 600 c/c with a 50 × 100 Keruing header to 50 visual grade.

It should not go unnoticed that the supported roof framing is also limited in bearing across the header. This would indicate that a thickness in excess of 50 mm is required if this framework is to be in Redwood.

Finally, when softwoods to Tables 3.1 and 3.2 are 65 grade and above, or when columns are formed from hardwoods, K_{19} becomes the modification factor and not K_{18}.

10.1.6 Axially loaded spaced columns

A spaced column can be described as two or more equal rectangular section shafts separated by spacing blocks positioned at each end and intermediate points. The intermediate spacers are normally positioned to achieve equidistant centre to centre disposition and either glued, bolted, screwed or nailed connected to the shafts. Fig. 10.13 indicates the typical make up of a spaced column.

The advantages of the use of spaced columns are to be found in such areas as triangulated frameworks and columns which carry heavy loads or which, by their use, afford an easy member to member connecting joint at the top seating.

When subjected to an axial compression load, the column has three limiting buckling planes which are X–X, Y–Y and S–S. These relate respectively to the major axis and minor axis of the composite unit and to the minor axis of the individual shaft. The largest slenderness ratio of these three axes will control the load carrying capacity of the spaced column. CP 112 stipulates several limitations for the purposes of design which may be listed and discussed as follows:

(a) Clear distance between shafts $\not> 3 \times$ thickness of shaft.

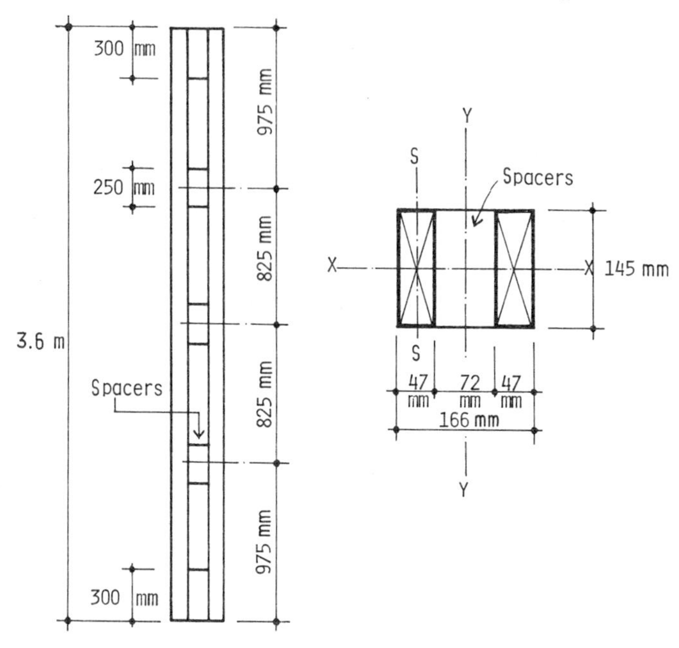

Fig. 10.13

(b) End packs – mechanical and glued connections ≮6 × thickness of shaft and sufficient to transmit a shearforce between faces of shaft and pack equal to

$$\frac{1.5 \times \text{total applied compressive load on whole column}}{\text{number of individual shafts}}$$

and fixed by cramping or at least two bolts or four screws spaced to give even pressure.

(c) Intermediate packs – glue ≮230 mm long and fixed by cramping or at least two bolts or four screws spaced to give even pressure.

(d) Intermediate packs – general requirements are that they should be spaced so that there are at least two packs in the length of the column or one if the column's length does not exceed 30 × thickness of shaft but, in any event, sufficient packs to ensure that the l/r of the axis S–S between packs is limited to 70 or 0.7 times the greatest l/r of the whole column, whichever is the least. The effective length of the individual shaft for this purpose should be taken as the centre to centre spacing of the packs.

(e) Load capacity for S–S is taken for the effective length between centre to centre packs calculated as for a rectangular column whose section is that of one shaft and multiplied by the total number of shafts.

(f) Load capacity for X–X is assessed for an effective length as described in Section 10.1 with r_x taken as for an individual shaft multiplied by the number of shafts.

(g) Load capacity for Y–Y is based again upon an effective length as described in Section 10.1 but which is increased by a modification factor K_{20} as given in Table 20 of CP 112. This factor allows for the method of connection between shafts and packs based upon the ratio of the space between the shafts to the shaft thickness.

Like solid section columns, the design of spaced columns is one of trial and error and in the example to follow we will see exactly how a given column is checked for design in accordance with the requirements of CP 112.

Example 10.3

The spaced column detailed in Fig. 10.13 is constructed from GS grade Whitewood and contains two 47 × 145 shafts spaced 72 mm apart. All packs are glue fixed to the shafts in the factory by clamping. Determine the maximum load carrying capacity of the column assuming long term loading.

Step 1 Check the geometric limitations imposed by the code and establish the maximum slenderness ratio for each axis and thereby establish the limit for the column.

Step 2 Determine K_{18} and hence the load carrying capacity of the column.

Step 3 Knowing the load carrying capacity check the shear force limitation in the end packs.

Step 1

The ratio between spacing and shafts = 72/47 = 1.53 < 3 Satisfactory
Ratio of end pack length to shaft thickness = 300/47 = 6.4, which is greater than the minimum limitation of 6 and so is satisfactory. Intermediate pack lengths to be minimum of 230 mm to comply with CP 112 < 250 provided and is satisfactory.

Slenderness ratio X–X:

Effective length = 3600 mm

r_x for individual 47 × 145 = 41.9 mm

giving

$$\frac{l}{r} = \frac{3600}{41.9} = 86$$

Slenderness ratio Y–Y:

Effective length = 3600 mm

$$r_y = \sqrt{\left(\frac{I_y}{A_y}\right)}$$

where

$$I_y = \frac{14.5}{12}(16.6^3 - 7.2^3) = 5076\ \text{cm}^4$$

$$A_y = 4.7 \times 14.5 \times 2 = 136.3\ \text{cm}^2$$

$$r_y = \sqrt{\frac{5076}{136.3}} = 6.1\ \text{cm}$$

$$\text{hence}\ \frac{l}{r} = \frac{3600}{61} = 59$$

Now increase by K_{20} which from Table 20 of CP 112 for glued connections and a ratio of 1.53 is given as 1.206 by interpolation. Finally,

$$\frac{l}{r} = 59 \times 1.53 = 90.3$$

Slenderness ratio S–S:

Effective length = 825 mm centres of packs

r_y for individual 47 × 145 = 13.6 mm

giving

$$\frac{l}{r} = \frac{825}{13.6} = 60.7$$

Now check spacing of packs against

$$\frac{l}{r}(\text{S–S}) \not> 70 \text{ or } \not> 0.7 \times \frac{l}{r}$$

greatest for column.
Greatest l/r = 90.3 giving limit of $90.3 \times 0.7 = 63.2$.
Therefore

$$\frac{l}{r}(\text{S–S}) = 60.7 < 63.2 \qquad \text{Satisfactory}$$

Hence the limiting l/r is that produced by the Y–Y axis and equals 90.3.

Step 2

From Table 15 of CP 112 the value of K_{18} for long term loading by interpolation for $l/r = 90.3$ is given as 0.607.

Now the load carrying capacity of the column is given by the expression

$$P = C_{\text{ppar}} \times A_s \times N_s$$

where

P = maximum permissible applied load

C_{ppar} = permissible compressive stress parallel to grain

A_s = area of individual shaft

N_s = number of shafts in the column

Hence

$$C_{\text{ppar}} = C_{\text{gpar}} \times K_{18}$$

$$= 5.6 \times 0.607 = 3.4\ \text{N/mm}^2$$

$$A_s = 4.7 \times 14.5 = 68.2\ \text{cm}^2$$

$$N_s = 2$$

giving

$$P = \frac{3.4}{10} \times 68.2 \times 2 = 46.3\ \text{kN}$$

Step 3

Shear force between faces of shaft and pack is given by the expression

$$V_f = 1.5 \frac{P}{N_s}$$

where V_f is the shear force between connecting faces. Hence

$$V_f = \frac{1.5 \times 46.3}{2} = 34.8\ \text{kN}$$

$$\text{shear stress } v = \frac{34\,800}{300 \times 145} = 0.800\ \text{N/mm}^2$$

permissible stress = grade stress parallel to grain

$$= 0.86\ \text{N/mm}^2 > 0.800 \text{ satisfactory.}$$

So, summarizing, the load carrying capacity of the given column is shown to be 46.3 kN.

Where spaced columns are subjected to axial compressive loads plus bending the method of design will be similar to that which follows in Section 10.2 for solid section posts. All that is required is to establish the bending stress about the particular axis under consideration based upon the total section modulus of that axis.

In triangulated frameworks such as trusses and girders, Clause 3.14.8 of CP 112 gives guidance as to effective lengths for the compression members and where these members are considered to be spaced columns all design should be in accordance with the design requirements for spaced columns except for the design of end connecting joints. In the case of these joints, the shear resistant requirement of Clause 3.14.6.1 may be disregarded.

10.2 Axially loaded posts plus bending

By now, the reader will know how to cope with the assessment of loading (both axial and wind), bending moments, shear and deflection. All of these are to be found in the design of a column subject to bending and axial load.

Part of the calculations, leading to the total solution, have already been dealt with in Section 10.1 and there remains only that part which examines the stresses imposed by the bending. Clauses 3.14.3 and 3.15.2 of CP 112:Part 2:1971 give the rules relating to the manner in which combined bending plus compression and bending plus tension are to be dealt with. In both cases, one must examine the sum of the stress ratios and equate to either 0.9 (compression) or 1.0 (tension). It is permissible to equate to 1.0 for bending plus compression when the l/r ratio does not exceed 20. However, a ratio as low as this is most unusual.

In regard to deflection, there is a body of opinion among engineers, which says that deflection caused by the action of the wind alone should be ignored. The reasoning is that wind is of short term duration and, as it can be shown that timber does not reach its full deflection for some hours, then a short term loading will have little detrimental effect where deflection is concerned. For the most part, the deflection of the column should not concern the designer unduly, though consideration must be given to the type of external cladding, internal finishes and the visual effect. In short, it is a question for individual assessment and possible discourse with the approving authority. One should also not be unmindful of the possible effects of deflection upon glazing and jointing techniques. Suffice to say that the most important feature is the strength of the column and, provided the summation of the stress ratio is satisfied, this is automatically taken care of.

Another point in column design, where bending stresses are induced, is that for the full bending stress to be allowed the breadth to depth ratios in Table 17 of the code must prevail. Normally, in solid section beams, this does not present a problem; however, in columns the wind can cause a reversal of

Table 17. It is, however, unusual because examination of Table 17 shows that an isolated column would normally have its ends held in position and so a ratio of 3 applies which is normally more than adequate. In the event that buckling is critical, there is good reference to a design method in Chapter 9 of the *Timber Designer's Manual* by Ozelton and Baird but this subject will not be entered into here.

Let us now examine the actual design and see what steps are taken in the calculations leading to the solution. As in the case of axial loading alone, the manner in which one arrives at the section size is largely trial and error.

Example 10.4

A 50 × 100 actual size section is subjected to a maximum bending moment induced by wind of 0.400 kN m and has an applied medium term axial load of 7.0 kN. It forms the stud in a load bearing external wall and it may be considered to be fully restrained in the directions of its weak axis. The post's full length is 2.4 m and the reduction factor for restraint about the X–X axis may be taken as 0.85. Check that this section satisfies the code of practice in regard to the requirements of Clause 3.14.3 given that the timber species is to be European Whitewood to M50 grade.

First follow Steps 1 to 3 as given in Example 10.1.

Properties: $I_{XX} = 5.0 \times \frac{10^3}{12} = 416.7 \text{ cm}^3 \qquad A = 50 \text{ cm}^2$

$$r_x = \sqrt{\frac{416.7}{50}} = 2.89 \text{ cm}$$

$$\text{Eff. } l = 2400 \times 0.85 = 2040 \text{ mm}$$

$$\frac{l}{r_x} = \frac{2040}{28.9} = 70.6 \text{ giving } K_{18} = 0.89 \text{ (med. term)}$$

$$C_{\text{gpar}} \text{ (M50) Table 3.11} = 7.1 \text{ N/mm}^2$$

$$C_{\text{ppar}} = 7.1 \times 0.89 \times 1.1 = 6.95 \text{ N/mm}^2$$

Step 4

Determine the applied compressive stress.

$$C_{\text{apar}} = \frac{70}{50} = 1.40 \text{ N/mm}^2$$

Step 5

Find the applied and permissible bending stresses.

$$f_{\text{apar}} = \frac{M}{Z_x} \quad \text{with } Z_x = 5.0 \times \frac{10^2}{6} = 83.3\ \text{cm}^3$$

$$= \frac{400}{83.3} = 4.8\ \text{N/mm}^2$$

$$f_{\text{ppar}} = 6.6 \times 1.1 \times 1.50\ \text{(short term)}$$

$$= 10.89\ \text{N/mm}^2$$

Step 6

Check the summation of stress ratios for compliance with Clause 3.14.3.

$$\text{Summations} = \frac{f_{\text{apar}}}{f_{\text{ppar}}} + \frac{C_{\text{apar}}}{C_{\text{ppar}}}$$

$$= \frac{4.8}{10.89} + \frac{1.40}{6.95}$$

$$= 0.440 + 0.201$$

$$= 0.641 < 0.90 \qquad \text{Satisfactory}$$

Example 10.5

The following line diagram (Fig. 10.14) indicates how an external column in an industrial building is loaded. One external cladding rail occurs at the mid-height of the column and the column's width must not exceed 75 mm. The

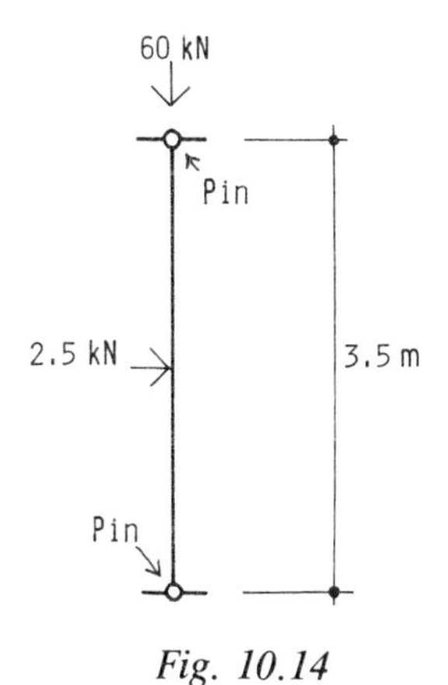

Fig. 10.14

columns are spaced at 3.6 m centres. Find the minimum depth of section required and check the summation of stresses. The axial load is of long term duration and the lateral load P is created by wind action. The timber is to be 50 grade Keruing with the permissible stresses as per Table 3.2.

An examination of Table 17 of the code indicates a maximum permissible depth to breadth ratio of 4. However, as the question asks for the minimum size we will try a ratio of 3 first. Hence $D = 75 \times 3 = 225$ mm.

Properties: $A = 7.5 \times 22.5 = 168.7\ \text{cm}^2 \qquad r_x = 6.5\ \text{cm} \qquad r_y = 2.17\ \text{cm}$

$$\frac{l}{r_x} = \frac{3500}{65} = 54 \qquad \frac{l}{r_y} = \frac{3500}{21.7 \times 2} = 81$$

Hence l/r_y governs, and for hardwoods K_{19} is the controlling modification factor.

In this case, the limiting value of l/r must first be compared against the formula

$$\sqrt{\frac{11.46E}{C_{\text{gpar}}}}$$

where E is the minimum value.

$$E = 9300\ \text{N/mm}^2 \text{ and } C_{\text{gpar}} = 8.3\ \text{N/mm}^2$$

Hence

$$\sqrt{\left(\frac{11.46 \times 9300}{8.3}\right)} = 113.3 > 81$$

this result indicates that the formula

$$1.00 - 0.0437\frac{C_{\text{gpar}}}{E}\frac{(l)^2}{(r)^2}$$

is to be used to find K_{19}, therefore substituting we have:

$$1.00 - 0.0437 \times \frac{8.3}{9300} \times 81^2 = 0.744$$

giving a

$$C_{\text{ppar}} = 8.3 \times 0.744 = 6.17\ \text{N/mm}^2$$

Note that the load sharing increase factor does not apply. Now

$$C_{\text{apar}} = \frac{600}{168.7} = 3.56\ \text{N/mm}^2$$

The next step is to determine the applied and permissible bending stresses, but first applied moment must be calculated

$$M = \frac{PL}{4} = 2.5 \times \frac{3.5}{4} = 2.19 \text{ kN m}$$

giving

$$f_{apar} = \frac{2190}{633} = 3.46 \text{ N/mm}^2$$

$$f_{ppar} = 9.7 \times 1.50 = 14.55 \text{ N/mm}^2$$

Finally the summation of stresses gives:

$$\frac{f_{apar}}{f_{ppar}} + \frac{C_{apar}}{C_{ppar}}$$

$$\frac{3.46}{14.55} + \frac{3.56}{6.17}$$

$$0.238 + 0.577 = 0.815 < 0.90 \qquad \text{Satisfactory}$$

Therefore the minimum section will be 75×225 mm. A check on the deflection gives:

$$d = \frac{PL^3}{48EI} = \frac{2500 \times 3500^3}{48 \times 9300 \times 71.2 \times 10^6}$$

$$= 3.37 \text{ mm}$$

It can be seen that even using $d_p = 0.003L$, the applied deflection is well inside the permissible limits.

10.3 Design of FINK profile truss

There are many differing shapes to timber trusses, most of which in this country are used in roof construction and, primarily, roofs of domestic dwellings. Of all the trusses available, the most common configurations in use are those of the FINK and FAN trusses, whose shapes are indicated in Fig. 10.15.

The code of practice which controls the structural design of trusses is CP 112:Part 3:1973 'Trussed rafters for roofs of dwellings'. Contained within this code are maximum permissible span tables for various chord sizes and roof slopes which relate solely to FINK and FAN trusses.

These span tables were obtained from many hundreds of prototype load tests carried out in accordance with the test procedures given in CP 112:Part

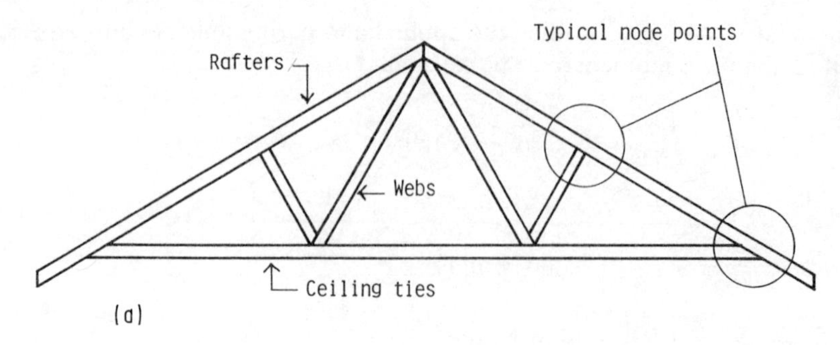

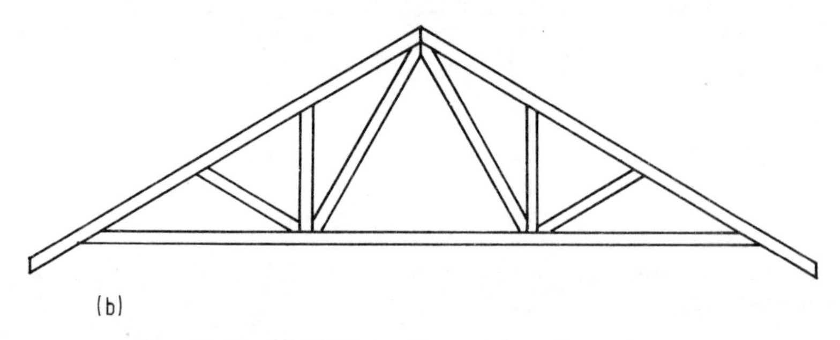

Fig. 10.15 (a) FINK profile truss (b) FAN profile truss

2:1971 and it is probably true to say that, because of this, timber roof trusses are the most deeply researched subject in the building industry.

Part 3 of the code also gives design parameters for arriving at the timber chord and web sizes, given that the span tables are not applicable to a particular requirement. It should, however, be noted that these parameters cannot be used in an attempt to justify the timber sizes given in the code. The span table limitations, having been arrived at by the expedience of load testing, require no further justification.

A close examination of Clause 5.3 shows that various load conditions are outlined and 50% continuity is allowed at the apex and heel (eaves) joints when arriving at the moments to be used in the design of the chord members. By utilizing this 50% fixity, it is possible to arrive at the magnitude of the moments occurring at the mid-span of the bays and at the node points (a 'node point' is simply any joint occurring around the perimeter of the truss).

For the case of the FINK truss, which we are to examine, these moments are best indicated for the rafter and ceiling tie chords, in the moment diagrams shown in Fig. 10.16. They have been obtained by using a moment distribution approach to both rafters and ceiling ties.

The design of the various members within the truss is a combination of

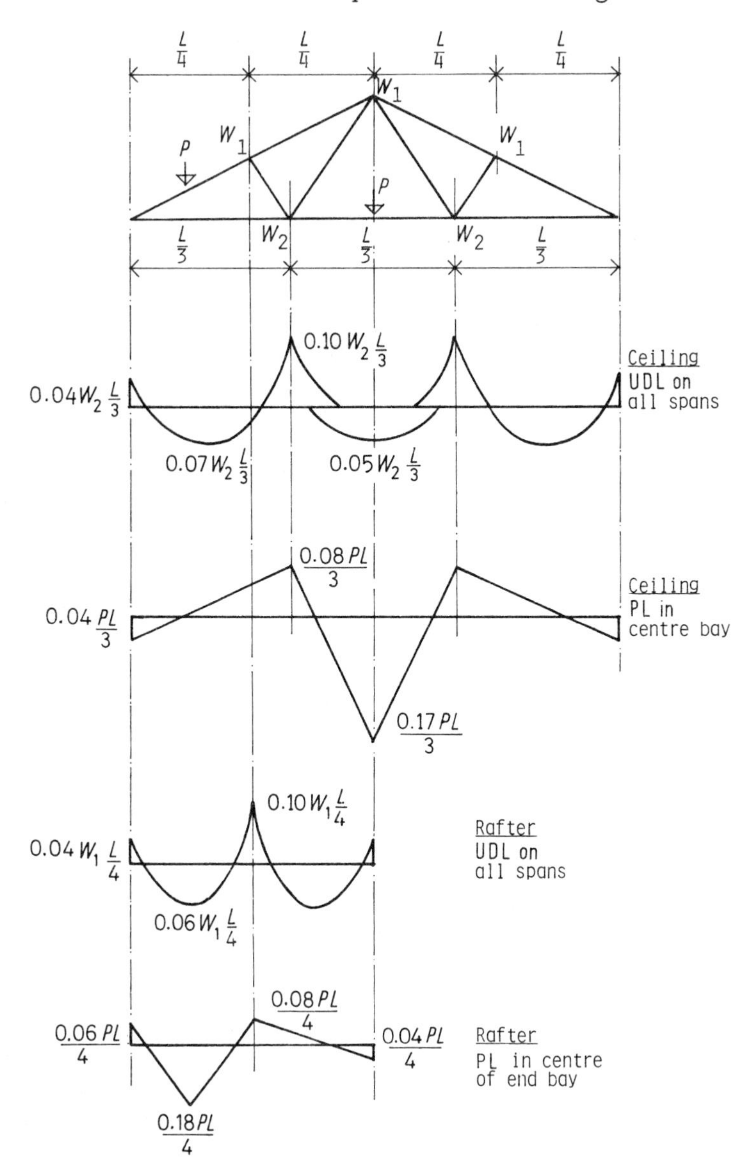

Fig. 10.16 Bending moment coefficients for FINK profile truss

bending and compression (rafters), bending and tension (ceiling ties) and direct axial compression or direct axial tension (webs). The principles of the structural design approach to these combinations have been covered in the earlier chapters and so the design of a truss is well within the capabilities of the reader having reached this far.

There only remains the solution to obtaining the axial forces in the members for the picture to be complete. In this respect, the most widely used method of approach among designers is to construct a 'force diagram'. Therefore, we will examine how this is done in the particular case of the FINK truss.

Example 10.6

Determine the axial loads in a FINK profile truss of 8.4 m overall span of wall plates, 30° slope and spaced at a maximum of 600 mm c/c. All design to be in accordance with CP 112:Part 3:1973, including standard applied loading as per Clause 3.0.

Step 1

Construct a line diagram of the truss, indicating the positions of the node loads, and apply a suitable notation for ease of reference to each member. The method used in Fig. 10.17 is known as 'bows notation'.

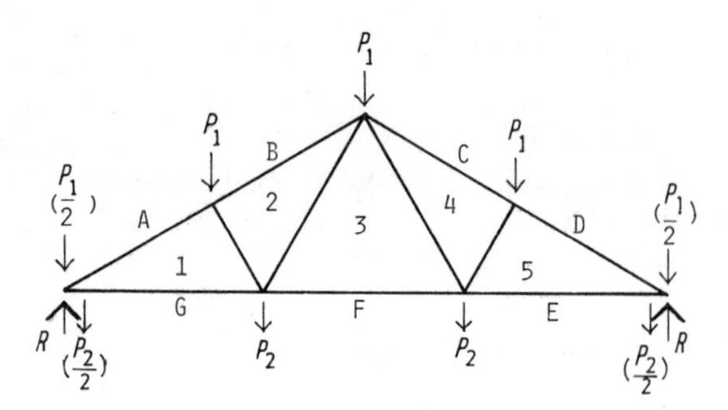

Fig. 10.17

Having constructed this diagram, it can be seen that the half node loads, acting at the supports, will have no effect upon the magnitude of the axial forces in the frame members. It is, therefore, possible to ignore this particular part of the loading and relate the final loading diagram to those parts (P_1 and P_2) which occur inside the reactions. However, it should not be overlooked that these half node loads do exist and must be added to the net reactions when checking for items such as bearing stress and total loads on supporting structure. A good procedure is to indicate these loads in parentheses so that the designer is reminded of their existence but does not include them in the force diagram build up.

Another small but significant point which is often overlooked is to include for the load induced by the eaves overhang (if any) in the total reaction build up.

Step 2

Determine the magnitude of the node loads P_1 and P_2 and draw the final loaded line diagram.

General

(a) Super (CP 3, Chapter V, Part 1:1967) = 0.75
(b) Dead (rafter) 0.685/cos 30° (Clause 3.0) = 0.79
(c) (Ceiling) Dead + Super = 0.50

(d) Total UDL loading = 2.04 kN/m^2
(e) In addition a concentrated load of 0.9 kN is applied to one of the ceiling tie node points to satisfy Clause 3.0

For loading (b) the figure of 0.685 kN/m^2 acts normal to the slope and it is necessary to divide this by the cosine of the angle of slope in order to convert it to a load acting on plan. In (c) the ceiling superimposed load of 0.25 kN/m^2 is added to the equivalent dead load because it may only be considered as long term loading.

For point (e) one must first consider if the 1.2 m headroom rule applies; if so, then the 0.9 kN concentrated load must be considered. With regard to its application on the rafters, it is difficult to see how it can influence the design of rafters when, very obviously, the tiles and tiling battens (unlike the ceiling) will distribute and consequently dilute the load to something appreciably less than 0.9 kN. In practice, this loading is generally ignored in rafter design; however, some designers would take it into consideration where the rafters were less than 97 mm in depth.

From this general loading may be found P_1 and P_2 as follows:

$$P_1 = (0.75 + 0.79) \times \frac{8.4}{4} \times 0.6 = 1.94 \text{ kN}$$

$$P_2 = 0.50 \times \frac{8.4}{3} \times 0.6 = 0.84 \text{ kN}$$

Therefore

$$R = 1.94 \times 1\tfrac{1}{2} + 0.84 = 3.75 \text{ kN}$$

and so the final loaded line diagram will be as indicated in Fig. 10.18.

Step 3

Construct a force diagram for the final load diagram and, by so doing, determine the axial loads in each member and indicate whether they are compressive or tensile.

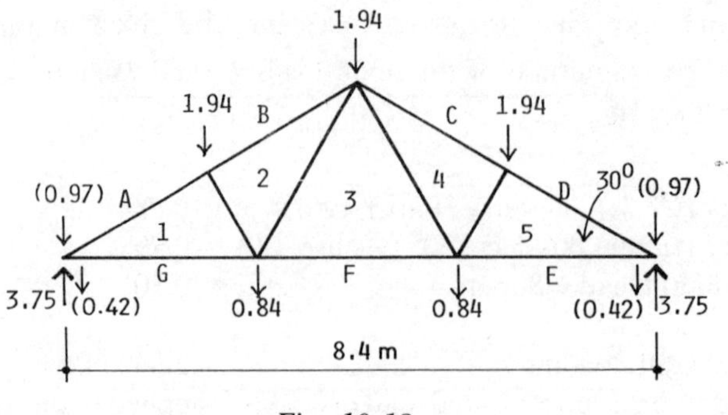

Fig. 10.18

A force diagram must always be drawn as accurately as possible and always to a chosen scale; the larger the scale, the more accurate will be the answer. On completion, by using this scale, the forces in each member can be arrived at. We will choose a scale of 1 cm = 1.0 kN and show that the final force diagram for a FINK truss is to the configuration in Fig. 10.19.

Because the truss and loading are symmetrical about the apex line, it is only necessary to construct that part of the diagram which is shown in heavy outline. It should be noted that the form of this diagram is common to all FINK profile trusses under symmetrical load and only varies in relation to the slope of the truss.

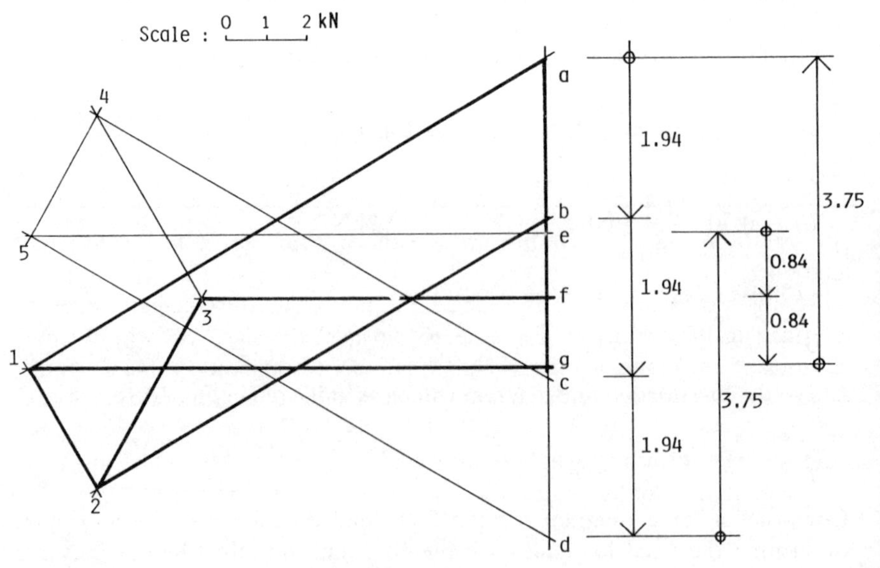

Fig. 10.19

By applying the scale, the axial loads may be given in the following tabulated form:

Member	*Axial load (kN)*
A1	−7.5
B2	−6.5
F3	+4.4
G1	+6.5
1/2	−1.7
2/3	+2.7

The − ve sign indicates compression whilst the + ve sign indicates tension.

We have now found the axial loads which are applicable to the UDL loading, but in addition to these are those created by the application of the 0.9 kN PL to the ceiling tie, because the 1.2 m headroom rule applies to this truss. By following the same procedure as before we can arrive at the values indicated in Fig. 10.20.

Once again the form of this diagram is common to all FINK trusses where the ceiling tie node point carries the 0.9 kN concentrated load and varies only with the slope.

The total tabulated axial loads may now be shown in the following cumulative way for the maximum conditions:

	Axial loads (kN)		
Member	*UDL*	*PL*	*Total*
A1	−7.50	−1.20	−8.70
B2 (A2)	−6.50	−1.20	−7.70
F3 (B3)	+4.40	+ .52	+4.92
G1 (C1)	+6.50	+1.04	+7.54
1/2	−1.70	—	−1.70
2/3	+2.70	+1.04	+3.74

The notations given in parentheses are those which apply to the PL line diagram.

Having now obtained the axial loads, we may proceed to the next stage of determining the timber section sizes to satisfy the design. We will use the previous example and assume M50 grade timber throughout (Table 3.11), in European Whitewood, with 35 × 97 top chords and 35 × 120 bottom chords and 35 × 72 webs.

Because the PL is applicable to this truss, all members should be justified for two conditions of loading:

Condition 1 Dead + Super (medium term)

Condition 2 Dead + Super + PL (short term)

Experience will show that these two conditions are normally the only ones which need to be checked and, indeed, for the ceiling tie, Condition 2 invariably governs.

We will, however, check both conditions when designing each of the principal members of the truss. For this truss, these principal members are A1 (rafter), F3 (ceiling tie), 1/2 (web strut) and 2/3 (web tie).

Rafter A1

In the case of the rafter, two positions need to be checked, the first being at the mid-point of the span and the second being at the node point formed by the junction of the rafter with the web strut.

Mid-point (Condition 1): The member is subject to axial compression plus bending and so Clause 3.14.3 of CP 112:Part 2:1971 applies.

$$\text{Axial load} = 7.50 \text{ kN}$$

$$\text{Moment} = 1.94 \times 2.1 \times 0.06 = 0.244 \text{ kN m}$$

$$\text{Properties: } A = 34 \text{ cm}^2 \qquad Z_x = 54.9 \text{ cm}^3 \qquad r_x = 28 \text{ cm}$$

$$\text{Eff. } l = \frac{2100}{\cos 30°} \times 0.80 = 1940 \text{ mm}$$

Therefore

$$\frac{l}{r_x} = \frac{1940}{28} = 70$$

hence

$$K_{18} = 0.90 \text{ (med. term)}$$

$$C_{\text{gpar}} = 7.1 \text{ N mm}^{-2}$$

$$C_{\text{ppar}} = 7.1 \times 0.90 \times 1.1 = 7.03 \text{ N/mm}^2$$

$$C_{\text{apar}} = \frac{75}{34} = 2.21 \text{ N/mm}^2$$

$$f_{\text{apar}} = \frac{244}{54.9} = 4.44 \text{ N/mm}^2$$

$$f_{\text{gpar}} = 6.6 \text{ N/mm}^2$$

$$f_{\text{ppar}} = 6.6 \times 1.1 \times 1.25 = 9.075 \text{ N/mm}^2$$

Summation of stresses:

$$\frac{4.44}{9.075} + \frac{2.21}{7.03} = 0.489 + 0.314 = 0.803 < 0.900 \qquad \text{OK}$$

Mid-point (Condition 2):

Axial load = 8.70 kN

Moment (as before) = 0.244 kN m

All properties as before but $K_{18} = 1.01$ (short term)

$$C_{\text{ppar}} = 7.1 \times 1.01 \times 1.1 = 7.88\ \text{N/mm}^2$$

The applied and permissible bending stresses are unaltered because the external loading remains the same.

$$C_{\text{apar}} = \frac{87}{34} = 2.56$$

Summation of stress:

$$0.489 + \frac{2.56}{7.88} = 0.489 + 0.325 = 0.814 < 0.900 \qquad \text{OK}$$

Therefore the rafter is satisfactory for both conditions at the mid-point of the span.

Node point (Condition 1):

Axial load = 7.50 kN

Moment = 1.94 × 2.1 × 0.10 = 0.407 kN m

In the case of the node point, the code permits the use of the full term factor with no reduction for slenderness ratio and equates the sum of the stress ratios to unity.

$$C_{\text{ppar}} = 7.1 \times 1.1 \times 1.25 = 9.76\ \text{N/mm}^2$$

$$C_{\text{apar}} = \frac{75}{34} = 2.21\ \text{N/mm}^2$$

$$f_{\text{apar}} = \frac{407}{54.9} = 7.41\ \text{N/mm}^2$$

$$f_{\text{ppar}} = 6.6 \times 1.1 \times 1.25 = 9.075\ \text{N/mm}^2$$

Summation of stress:

$$\frac{7.41}{9.075} + \frac{2.21}{9.76} = 0.816 + 0.226 = 1.042 > 1.00$$

Node point (Condition 2):

$$\text{Axial load} = 8.70\text{ kN}$$

$$\text{Moment (as before)} = 0.407\text{ kN m}$$

$$C_{\text{apar}} = \frac{87}{34} = 2.56\text{ N/mm}^2$$

$$C_{\text{ppar}} = 7.1 \times 1.50 \times 1.1 = 11.71\text{ N/mm}^2$$

Summation of stresses:

$$0.816 + \frac{2.56}{11.71} = 0.816 + 0.219 = 1.035 > 1.0$$

In both cases, the summation factors exceed unity but are within 5% and so it could be argued that the rafter size is acceptable. If this did not prove to be acceptable to the checking authority then it can be seen, without further calculation, that the next size will work.

Ceiling tie F3

Like the rafter, the ceiling tie needs to be checked first at the mid-span and second at the node point.

Mid-span (Condition 1): This member is subject to axial tension plus bending and so Clause 3.15.2 of CP 112: Part 2:1971 applies.

$$\text{Axial load} = 4.4\text{ kN}$$

$$\text{Moment} = 0.84 \times 2.8 \times 0.05 = 0.118\text{ kN m}$$

$$\text{Properties: } A = 42\text{ cm}^2 \qquad Z_x = 8.40\text{ cm}^3$$

$$t_{\text{gpar}} = 4.6\text{ N/mm}^2$$

$$t_{\text{ppar}} = 4.6 \times 1.25 \times 1.1 = 6.32\text{ N/mm}^2$$

$$t_{\text{apar}} = \frac{44}{42} = 1.05\text{ N/mm}^2$$

$$f_{\text{ppar}} = 6.6 \times 1.00\text{ (long term)} = 6.60\text{ N/mm}^2$$

(load sharing is not permitted on bending stresses in the ceiling tie, unless there is adequate lateral distribution of the load)

$$f_{\text{apar}} = \frac{118}{84} = 1.404\text{ N/mm}^2$$

Summation of stresses:

$$\frac{1.404}{6.6} + \frac{1.05}{6.32} = 0.213 + 0.166 = 0.379 < 1.0 \qquad \text{OK}$$

Mid-span (Condition 2):

$$\text{Axial load} = 4.92\ \text{kN}$$

$$\text{Moment} = \text{UDL moment} + \text{PL moment}$$

$$= 0.118 + 0.90 \times 2.8 \times 0.17 = 0.546\ \text{kN m}$$

$$t_{\text{ppar}} = 4.6 \times 1.50 \times 1.1 = 7.59\ \text{N/mm}^2$$

$$t_{\text{apar}} = \frac{49.2}{42} = 1.17\ \text{N/mm}^2$$

$$f_{\text{ppar}} = 6.6 \times 1.50 = 9.90\ \text{N/mm}^2$$

$$f_{\text{apar}} = \frac{546}{84} = 6.5\ \text{N/mm}^2$$

Summation of stresses:

$$\frac{6.5}{9.9} + \frac{1.17}{7.59} = 0.657 + 0.154 = 0.811 < 1.0 \qquad \text{OK}$$

Node point (Condition 1):

$$\text{Axial load} = 6.50\ \text{kN}$$

$$\text{Moment} = 0.84 \times 2.8 \times 0.10 = 0.235\ \text{kN m}$$

$$t_{\text{ppar}}\ \text{(as before)} = 6.32\ \text{N/mm}^2$$

$$t_{\text{apar}} = \frac{65.0}{42} = 1.55\ \text{N/mm}^2$$

$$f_{\text{ppar}} = 6.6 \times 1.0 = 6.6\ \text{N/mm}^2$$

$$f_{\text{apar}} = \frac{235}{84} = 2.80\ \text{N/mm}^2$$

Summation of stresses:

$$\frac{1.55}{6.32} + \frac{2.80}{6.60} = 0.245 + 0.424 = 0.669 < 1.0 \qquad \text{OK}$$

Node point (Condition 2):

$$\text{Axial load} = 7.45 \text{ kN}$$

$$\text{Moment} = 0.235 + 0.9 \times 2.8 \times 0.08 = 0.437 \text{ kN m}$$

$$t_{\text{ppar}} = 4.6 \times 1.50 \times 1.1 = 7.59 \text{ N/mm}^2$$

$$t_{\text{apar}} = \frac{75.4}{42} = 1.80 \text{ N/mm}^2$$

$$f_{\text{ppar}} = 6.6 \times 1.50 = 9.90 \text{ N/mm}^2$$

$$f_{\text{apar}} = \frac{437}{84} = 5.20 \text{ N/mm}^2$$

Summation of stresses:

$$\frac{5.20}{9.90} + \frac{1.80}{7.59} = 0.525 + 0.237 = 0.762 < 1.0 \qquad \text{OK}$$

The ceiling tie is therefore proven to be satisfactory.

Web 1/2 (Condition 1 only)

By referring to the force diagram, Fig. 10.19, it can be seen that the strut webs 1/2 and 4/5 are not affected by the point load, there being no additional load transmitted. They are, in effect, redundant as far as this load condition is concerned, in which case only Condition 1 applies.

$$\text{Axial load} = 1.70 \text{ kN}$$

$$\text{Properties: } A = 25.2 \text{ cm}^2 \qquad r_y = 10.1 \text{ mm}$$

$$\text{Eff. } l = 1400 \times 0.90 = 1260 \text{ mm}$$

$$\frac{l}{r_y} = \frac{1260}{10.1} = 125 \quad \text{hence} \quad K_{18} = 0.395 \text{ (med. term)}$$

$$C_{\text{ppar}} = 7.1 \times 1.1 \times 0.395 = 3.08 \text{ N/mm}^2$$

$$C_{\text{apar}} = \frac{17}{25.2} = 0.68 \text{ N/mm}^2 < 3.08 \qquad \text{OK}$$

Web 2/3 (Condition 1)

Subject to direct tension only:

$$\text{Axial load} = 2.70 \text{ kN}$$

$$t_{\text{ppar}} = 4.6 \times 1.1 \times 1.25 = 6.32 \text{ N/mm}^2$$

$$t_{\text{apar}} = \frac{27}{25.2} = 1.07 \text{ N/mm}^2 < 6.32 \qquad \text{OK}$$

(Condition 2):

Axial load $= 3.74$ kN

$$t_{ppar} = 4.6 \times 1.1 \times 1.50 = 7.59 \text{ N/mm}^2$$

$$t_{apar} = \frac{37.4}{25.2} = 1.49 \text{ N/mm}^2 < 7.59 \qquad \text{OK}$$

The webs are therefore proven to be satisfactory.

The final stage in the design of any truss should be the proof of the actual deflection when compared to the code of practice. Unfortunately, it is a fact that many trussed rafters are never checked for their deflection performance and, strangely, many checking authorities never think to question this criterion. Yet the code is quite specific and lays down definitive guidelines on the subject in that the long term deflection should be investigated at both the ceiling tie node point and the centre span of the ceiling tie mid-bay member. Clause 5.4 of CP 112:Part 3:1973 governs.

The deflection of a truss is caused by the cumulative effects of members shortening under compression, lengthening under tension and joints rotating and slipping. Deflection may be determined by adopting the strain energy formula for members subjected to axial strain:

$$d_n = \sum \frac{PUL}{AE}$$

where

d_n = the elastic deflection at a selected node point.

P = the load in each member induced by the applied loading (in this case long term).

U = the load in each member induced by a unit load applied at the node point for which the deflection is required and in the direction of the deflection.

L = the length of members between the node points (i.e. the actual timber lengths not the theoretical line diagram lengths).

A = the actual area of each member.

E = the modulus of elasticity of each member, bearing in mind that very often webs may have different values to those of the chords (for trusses spaced at a maximum of 610 mm c/c $E = E_{mean}$).

Although, at first glance to the inexperienced, this formula and its solution may look a little complex, the reader should not be deterred. Nearly all of the criteria required have been determined by the previous calculations and all that is necessary is to present the information in a tabulated form in order to arrive at the solution. Therefore, if we examine the previous example, we find that the deflection is found as follows:

Member	*P* *(N)*	*U*	*L* *(mm)*	*A* *(mm²)*	*E* *(N/mm²)*	*PUL/AE* *(mm)*
A1	−4740	−1.33	2425	3400	9000	0.500
B2	−4110	−1.33	2425	3400	9000	0.433
C4	−4110	−0.67	2425	3400	9000	0.218
D5	−4740	−0.67	2425	3400	9000	0.252
E5	+4110	+0.58	2800	4200	9000	0.176
F3	+2782	+0.58	2800	4200	9000	0.120
G1	+4110	+1.16	2800	4200	9000	0.353
1.2	−1075	0.00	1280	2520	9000	0.000
2.3	+1707	+1.16	2650	2520	9000	0.232
3.4	+1707	+0.00	2650	2520	9000	0.000
4.5	−1075	0.00	1280	2520	9000	0.000
					d_n =	2.284 mm

Having determined the crucial elastic deflection d_n at the node point, we must now refer to Clause 5.4 to check that the actual requirements of the code are satisfied.

(1) Deflection at the ceiling tie node point
(2) Deflection at the centre of the ceiling tie

The first one is simply

$$d_{(1)} = d_n \times 1.75$$

$$= 2.484 \times 1.75 = 4.00 \text{ mm}$$

The normal limitation is 10 mm and so this truss is satisfactory at this point.

The second one is a little more involved but, nevertheless, reasonably straightforward:

$$d = 1.75\left(1.15d_n + \frac{WL^4 \times 10^7}{EI}\right) \text{ mm}$$

and in this particular case

$$L = 2.8 \text{ m}$$

$$W = 500 \times 0.60 = 300 \text{ N/m}$$

$$E = 9000 \text{ N/mm}^2$$

$$I = I_x = 5.04 \times 10^6 \text{ mm}^4$$

hence substituting

$$d_{(2)} = 1.75\left(1.15 \times 2.284 + \frac{300 \times 2.8^4 \times 10^7}{9000 \times 5.04 \times 10^6}\right)$$

$$= 11.71 \text{ mm}$$

Now the cumulative deflection over the whole span measured at the centre point of the ceiling line is

$$d_{(1)} + d_{(2)} = 4.00 + 11.71$$

$$= 15.71 \text{ mm}$$

To comply with non-cracking of finishes the limitation set is

$$0.003 \times \text{span} = 0.003 \times 8400$$

$$= 25.2 \text{ mm} > 15.71 \text{ mm}$$

Once again, this truss is satisfactory at this point.

Note: In (1) and (2) the factor of 1.75 is an empirical increase over the elastic deflection to allow for the effects of joint slip and rotation.

Although it has taken quite a number of pages to cover this subject, it will be seen, by cutting through the descriptive notes, that the design of a truss merely utilizes the skills previously learnt of designing beams and columns subject to both axial load and axial load plus bending moments.

There is a feature of truss configuration selection, often used by the unwitting, which can lead to deflection problems. In some instances, a FAN profile truss is selected over a FINK in order to gain economies in the rafter size. What is overlooked in doing this is the all-important design deflection.

If we examine the force diagram for the unit load (Fig. 10.20) it will be seen that it is precisely the same for both configurations. Consequently, by having

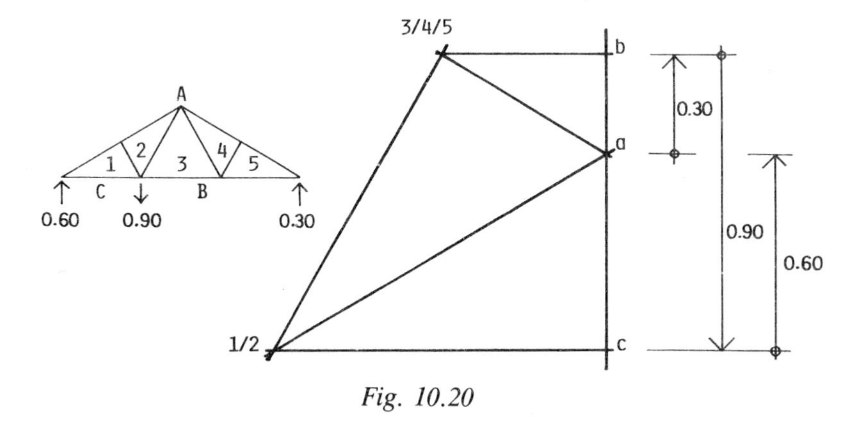

Fig. 10.20

reduced the area of the top chord, we have possibly increased the final deflection, probably to the detriment of the permissible deflection limitations. So a straightforward substitution is not necessarily correct, unless of course one is dealing with tested span tables, in which case the deflection is automatically checked.

Finally, most prefabricated timber trusses have their joints formed by the use of proprietary toothed-plate connectors. These connectors are discussed in depth in Chapter 12, Section 12.8.

11

Examples of element design – Part III

11.1 Design of composite tee-section

11.1.1 Introduction

As any student of structural engineering will realize, a simple change in shape of a cross-section can cause a considerable increase (or reduction) in its stiffness. If, for example, a piece of paper is held between finger and thumb at one end, the other end will immediately droop towards the ground. However, by simply folding the same piece in half and holding it at the root of the fold, such that it now forms a vee in cross-section, the resulting droop will be almost eliminated.

This is a very simple example but it serves to illustrate the significance of composite sections and how, by using common rectangular timber sections, their stiffness may be increased. Such is the case with tee-sections where two rectangular pieces are joined together in order to obtain a more efficient cross-section in a very economical manner.

Tee-sections may be used to resist bending, axially applied loads or a combination of both. If this form of section is to be analysed to perform a defined function, then it is usual to use a glued connection between the flange and the web (Fig. 11.1). Although mechanical means are possible, such as screws, nails and coach bolts, it is doubtful if the full potential of the section could be realized because of some of the limiting features of these mechanical

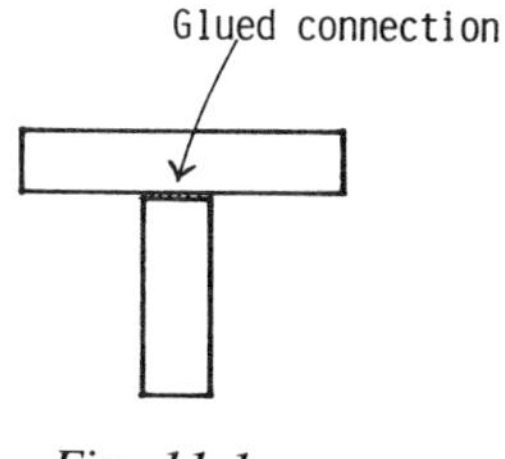

Fig. 11.1

fastenings, i.e. edge spacing, centre spacings, horizontal shear stresses, slip, etc.

Where mechanical means are used, it is generally in members where stresses are considered nominal and where good building practice dictates, rather than a true design analysis. A common example of this occurs frequently in roof truss designs where lateral stability of a web member may be achieved by nailing a horizontal piece to the web throughout its length thus changing its section to that of a tee.

11.1.2 Section properties

Before the full design of a tee-section can commence, it is necessary to know the cross-sectional properties. The properties required will of course depend upon the loads to be resisted and what the section is called upon to do. If, however, we assume that the conditions of bending and axial compression apply, then we would need to know area (A), first moment of area (Q), second moment of area (I), section modulus (Z) and radius of gyration (r).

In the case of a tee-section, we have two axes to consider: the stronger X–X and the weaker Y–Y but, before those properties about axis X–X may be

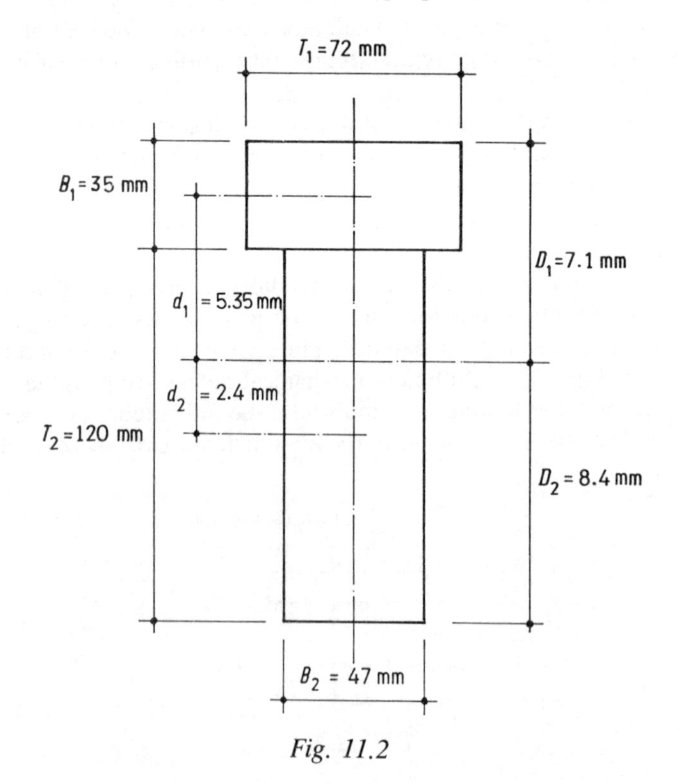

Fig. 11.2

determined, the centre of gravity of the section has to be found. However, it is not considered that further explanation of this particular procedure is necessary.

For those dimensions defined in Fig. 11.2 the corresponding properties about both axes are given as

$$A = B_1T_1 + B_2T_2$$
$$= 3.5 \times 7.2 + 4.7 \times 12.0$$
$$= 81.6 \text{ cm}^2$$

$$Q_x = 0.5D_2^2B_2$$
$$= 0.5 \times 8.4^2 \times 4.7$$
$$= 166 \text{ cm}^3$$

$$I_x = \frac{B_1^3T_1}{12} + B_1T_1d_1^2 + \frac{T_2^3B_2}{12} + B_2T_2d_2^2$$
$$= \frac{3.5^3}{12} \times 7.2 + 25.2 \times 5.35^2 + \frac{12^3}{12} \times 4.7 + 56.4 \times 2.4^2$$
$$= 1749 \text{ cm}^4$$

$$Z_x(\text{max.}) = \frac{I_x}{D_1}$$
$$= \frac{1749}{7.1}$$
$$= 246 \text{ cm}^3$$

$$Z_x(\text{min.}) = \frac{I_x}{D_2}$$
$$= \frac{1749}{8.4}$$
$$= 218 \text{ cm}^3$$

$$r_x = \sqrt{\frac{I_x}{A}}$$
$$= \sqrt{\frac{1749}{81.6}}$$
$$= 4.63 \text{ cm}$$

$$I_y = \frac{T_1^3 B_1}{12} \times \frac{B_2^3 T_2}{12}$$

$$= \frac{7.2^3}{12} \times 3.5 + \frac{4.3^3}{12} \times 12$$

$$= 213 \text{ cm}^4$$

$$Z_y = \frac{I_y \times 2}{T_1}$$

$$= \frac{213 \times 2}{7.2}$$

$$= 59.2 \text{ cm}^3$$

$$r_y = \sqrt{\frac{I_y}{A}}$$

$$= \sqrt{\frac{213}{81.6}}$$

$$= 1.62 \text{ cm}$$

To illustrate how these properties are applied, we will consider two examples. The first will be a beam carrying a uniformly distributed load and the second we will take as a column subject to axial load and bending with the flange in compression for both examples.

Example 11.1

The beam indicated in Fig. 11.3 is carrying a long term load of 4.0 kN. Check the bending and shear stresses assuming SS grade European Whitewood and check the deflections.

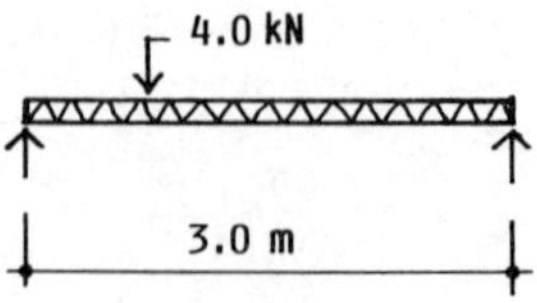

4.0 kN

3.0 m

Fig. 11.3

As the flange is in compression, it may be assumed that the full bending stress may be realized because of the shape of the section. If the lower section of the web were to be in compression, a quick solution is to check the web's breadth/depth ratio, as if acting alone, for compliance with Table 17 of CP 112:Part 2:1971. All properties of beam section as for Section 11.1.2.

$$M = 4.0 \times \frac{3.0}{8} = 1.50 \text{ kN m}$$

$$f_{\text{apar}} = \frac{M}{Z_x(\text{min.})}$$

$$= \frac{1500}{208} = 7.21 \text{ N/mm}^2$$

$$f_{\text{ppar}} = 7.3 \text{ N/mm}^2 \text{ (Table 3.7)}$$

Maximum shear $V = 2.00$ kN

$$v = \frac{VQ_x}{B_2 I_x}$$

$$= \frac{2000 \times 166}{47 \times 1749 \times 10} = 0.404 \text{ N/mm}^2$$

$$V_{\text{ppar}} = 0.86 \text{ N/mm}^2$$

To check the shear stress at the glue line, we must determine the first moment of area at the glue line which is:

$$Q \text{ (glue line)} = 4.7 \times 12.0 \times 2.4 = 135 \text{ cm}^3$$

Now

$$v \text{ (glue line)} = \frac{2000 \times 135}{47 \times 1749 \times 10} = 0.328 \text{ N/mm}^2$$

It is determined (Section 12.7.3) that, where the glue line interface is acting parallel to the grain, the full shear stress in the parent wood may be developed. In this case, we have a stress which is parallel to the grain in both members and so the applied stress is satisfactory.

It should be noted that if nail pressure glueing were to be used, then the permissible stress would need to be reduced by the factor of 0.9 (Section 12.7.3) giving a permissible stress of 0.774 N/mm^2. Finally the deflection is found:

$$d = \frac{5WL^3}{384EI_x} \quad \text{with } E = E_2 = 6960 \text{ N/mm}^2$$

$$= \frac{5 \times 2000 \times 3000^3}{384 \times 6960 \times 1749 \times 10^4} = 5.78 \text{ mm}$$

$$d_p = 0.003 \times 3000 = 9.0 \text{ mm} > 5.78 \qquad \text{Satisfactory}$$

Example 11.2

The post indicated in Fig. 11.4 is formed from the same tee-section and is subjected to an axially applied load in addition to bending caused by wind action. Take for the flange in compression and check the strength of the post in accordance with the requirements of CP 112:Part 2, Clause 3.14.3 and check the deflection.

$$P = 15.0\text{ kN (long term)}$$

$$W = 1.5\text{ kN (short term)}$$

$$L = 2.40\text{ m}$$

$$\frac{l}{r_x} = \frac{240}{4.63} = 51\text{ (assuming no end restraints)}$$

$$\frac{l}{r_y} = \frac{240 \times 0.85}{1.62} = 126\text{ (assuming partial end restraints)}$$

Consequently l/r_y governs the design.

$$K_{18} = 0.373\text{ (long term)}$$

$$C_{\text{ppar}} = 8.0 \times 0.373 = 2.98\text{ N/mm}^2$$

$$C_{\text{apar}} = \frac{150}{81.6} = 1.84\text{ N/mm}^2$$

$$f_{\text{ppar}} = 7.3 \times 1.50\text{ (short term)}$$

$$= 10.95\text{ N/mm}^2$$

$$M = 1.5 \times \frac{2.4}{8} = 0.45\text{ kN m}$$

$$f_{\text{apar}} = \frac{M}{Z\text{ (min.)}} = \frac{450}{208} = 2.16\text{ N/mm}^2$$

Summation of stress:

$$\frac{C_{\text{apar}}}{C_{\text{ppar}}} + \frac{f_{\text{apar}}}{f_{\text{ppar}}}$$

$$\frac{1.84}{2.98} + \frac{2.16}{10.95} = 0.815 < 0.90 \qquad \text{Satisfactory}$$

Maximum shear $V = 0.750$ kN

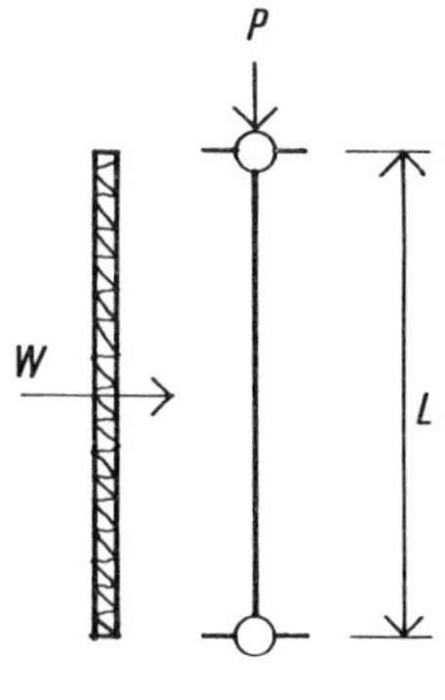

Fig. 11.4

By comparing this shear with the previous Example 11.1, it can be seen that there would be a smaller result and so no further justification is necessary.

For this particular case, if nail pressure were to be used, then

$$v_{ppar} = 0.86 \times 0.90 \times 1.50 = 1.161 \text{ N/mm}^2$$

The deflection will be

$$d = \frac{5 \times 1500 \times 2400^3}{384 \times 6960 \times 1749 \times 10^4} = 2.22 \text{ mm}$$

$$d_p = 0.003 \times 2400 = 7.2 \text{ mm} > 2.22 \qquad \text{Satisfactory}$$

11.1.3 Bending about Y–Y axis

In concluding this chapter, where bending should have to be considered in the plane of the Y–Y axis, then the additional sectional property of Q_{WW} (Fig. 11.5) would need to be calculated in order to determine the horizontal shear at this junction in the section.

Hence

$$Q_{WW} = bT_1y$$

Once again, should the area bT_1 prove to be unrestrained against buckling then a check for compliance with Table 17 of CP 112 will suffice.

11.2 Design of wind shear racking panels

11.2.1 Introduction

Panels placed in the vertical attitude which are called upon to resist lateral forces, generally transmitted through one of the upper corners, are known as

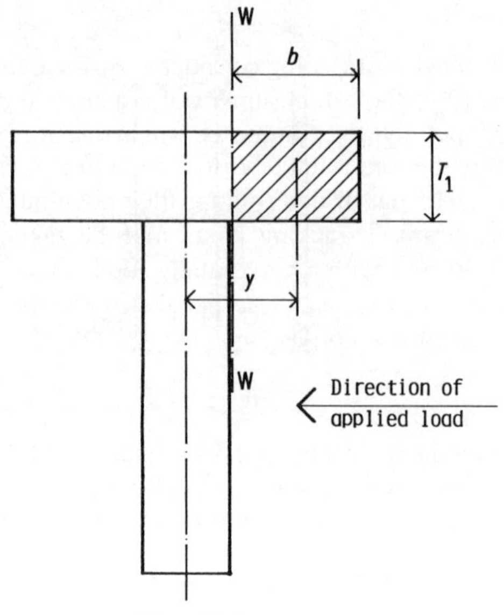

Fig. 11.5

'shear panels' or 'racking panels'. In the case of timber frame construction, the latter of these two names is more often referred to because the panels are normally used to resist the racking (lateral distortion) caused by the action of wind. The construction of a timber framed racking panel invariably consists of softwood vertical studs at 400–600 mm centre spacing backed with, more commonly, plywood sheathing. Other materials are available such as tempered hardboard and bitumen impregnated insulation board and, in the case of timber framed housing, the sheathing material is used on one side only; this being to the external face of external wall panels and to a choice of either face when using internal racking panels.

The fixing of the sheathing to the softwood frame is normally by means of nailing at various centres depending upon the type of sheathing used but these centres will generally be 75, 100 or 150 mm around the perimeter where the higher loads occur and at a maximum of 300 mm centres to the inner verticals. Staples can be used but, depending upon the leg diameter, they should be spaced at closer centres.

11.2.2 Design

The design of a wind shear racking panel may be justified by one of two methods:

(a) by theoretical analysis
(b) by making use of prototype test results

(a) Theoretical analysis

Currently there is much work being considered by the CSB/32/14 (BS 5268) committee in regard to the establishment of a definitive design method for racking panels. Hopefully, the committee will arrive at some method which will tend to simplify the somewhat laborious procedure involving the analysis of the perimeter nail fixings by determining their polar modulus.

In the simplest terms, a racking panel may be taken to be a vertical cantilever fixed at its base with a horizontally applied load at the upper edge (Fig. 11.6) or a combination of these panels fixed in line or separated by openings such as doors or windows.

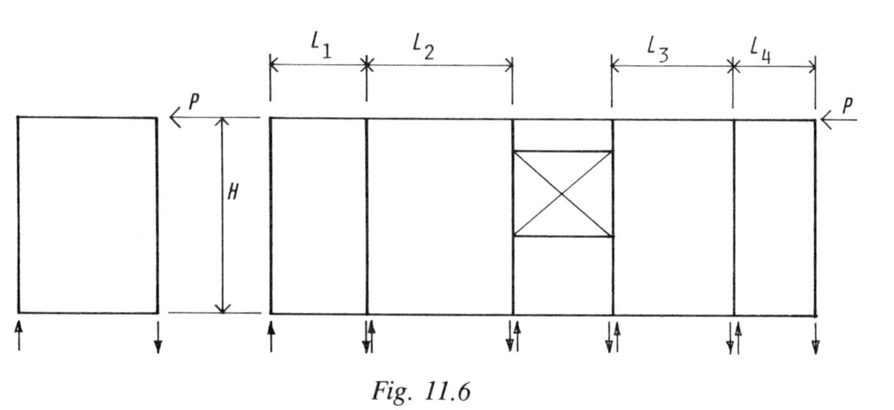

Fig. 11.6

If a racking load P is applied as indicated in Fig. 11.6 then in a combination of panels the amount of load resisted by each panel will be in proportion to its relative stiffness. This may be taken as L_1^2/H, L_2^2/H, etc. The greater the vertical load applied to the top of the panel the greater will be its resistance to the action of the horizontal load. Consequently, the worst condition for assessing the design limitations is when the minimum possible vertical load is applied. Sometimes, this load may be an uplift load occurring when wind suction overcomes the applied dead loads.

In all cases, it will be necessary to determine if the net overturning moment is causing an uplift effect at one corner, in which case vertical anchorage will be necessary. If we examine the loading conditions given in Fig. 11.7 we can see how this is derived by taking moments about the leading lower corner.

By taking moments about the point A the magnitude and direction of the reaction R may be evaluated:

$$R = \frac{2.0 \times 2.4 - 5.0 \times 1.2 - 5.0 \times 0.60}{1.2}$$

$$= -3.5 \text{ kN}$$

The negative sign indicates that R is acting in the same direction as the vertically applied loads and therefore uplift does not occur. In consequence, the only force which needs to be resisted is the horizontal shear along the bottom edge of 2.0 kN. Both the uplift and horizontal forces would be resisted by mechanical means such as bolts, nails and steel straps.

Within the construction of the panel, there exists the need to translate the horizontal force into the nails which connect the sheathing material to the vertical studs and upper and lower horizontal rails. The justification of these fixings can be arrived at by adopting the polar modulus method of approach.

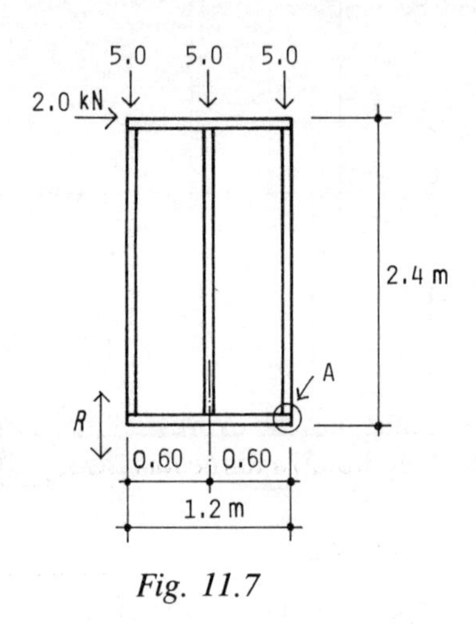

Fig. 11.7

It is, however, a very long and tedious method and rarely is it actually used in practice. In the panel indicated in Fig. 11.8 the applied racking load (R) tends to rotate the panel about the lower opposite corner and to distort the panel as indicated by the broken line profile.

This distortion is considered to be resisted by an internal couple set up by the resistance of each nail in relation to its stiffness and its distance from the centre of gravity of the panel, thus generating a polar moment of resistance.

In this case, the most heavily loaded nails will be those placed in the corners of the panel and the shear may be found by simply applying the standard formula of

$$M/Z$$

where

$M = Rh$

$Z =$ polar section modulus

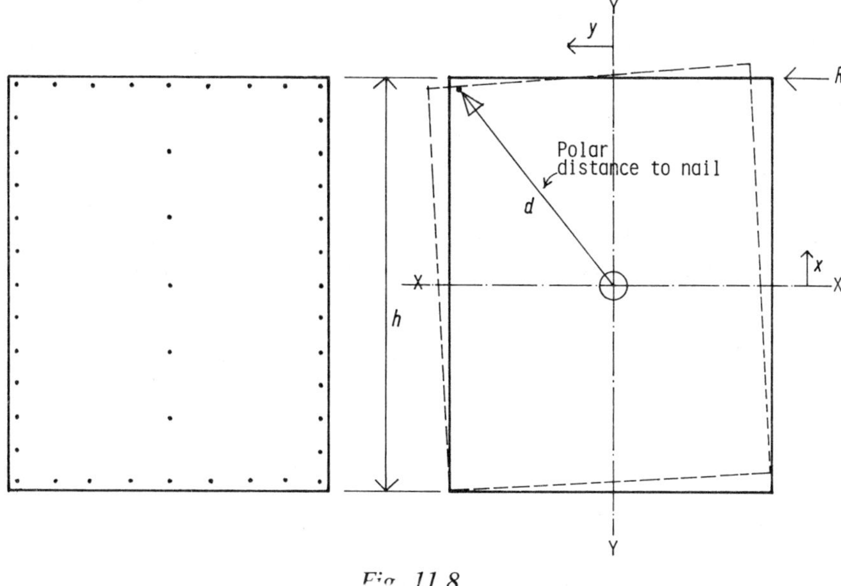

Fig. 11.8

The polar section modulus may be expressed in relation to the vertical and horizontal axes of X–X and Y–Y and the distance d to the further nail as

$$\frac{2(\Sigma x^2 + \Sigma y^2)}{d}$$

The values of x and y are the respective distances from the X–X and Y–Y axes to each nail giving the expression for the extreme nail shear

$$V_{ne} = \frac{M}{Z} = \frac{Rhd}{2(\Sigma x^2 + \Sigma y^2)}$$

However, even though the shear may be obtained in this manner, the effect of nail joint slip must be included together with the effects of sway and this may very well lead to the necessity for prototype testing.

Although we have very briefly touched upon a theoretical method of design, as previously stated, it is not commonly used and the normal course of events tends towards the use of information derived from test results which leads us into the next section.

(b) Prototype test results

In general, the commonest sheathing material used in racking panels is plywood and there has been, over a number of years, much testing of such panels. Normally, these tests relate to panels which do not exceed 2.4 m in height because this satisfies both the manufactured sheathing lengths and the domestic room height and so satisfies the majority of cases where wind racking panels are employed. However, there is no reason why greater height panels cannot be tested provided there is sufficient flexibility in the test rig equipment.

The design values are derived from a test procedure incorporating the application of variable vertical loads to each vertical stud in the panel and applying a horizontal load along the top rail to produce a deflection equivalent to $H/500$. Four cycles of load are applied for each condition of vertical load and the lowest racking load is taken as the test load. To allow for the correlation to a 50 year return wind condition, this test load is increased by multiplying by 1.25. If only one panel is tested, then a reduction factor of 0.8 is applied but if five panels are tested then the factor is taken as 1.0. The resulting figure is taken to be the design load and it must not exceed the ultimate load divided by a factor ranging from 2 if one panel is tested to 1.6 if the result is the lowest of five or more panels.

Such test results will, of course, relate to a particular specification but, under normal circumstances, for domestic design the plywood will be 9.5 mm or 12.5 mm and the nails will be 11 g × 50 mm, spaced at 150 mm around the perimeter members and 300 mm elsewhere. The test results should always be related to a magnitude of vertical load and be expressed as a permissible load per unit length, i.e. kN/m.

In Table 11.1 typical racking values are given for panels constructed from softwood framing at 600 mm c/c sheathed with 9.5 mm Douglas Fir plywood and fixed with 3.35 mm × 50 mm nails with spacing as previously discussed. The softwood framing is CLS Hem-Fir 38 × 89 in section.

The values in Table 11.1 are reproduced by kind permission of the Council of Forest Industries of British Columbia.

Table 11.1

0 kN per stud	2.5 kN per stud	5.0 kN per stud
2.2 kN/m	3.4 kN/m	3.9 kN/m

The ease of application of such test results to design problems is obvious and the consequential work is dramatically reduced. Other organizations such as TRADA produce similar tables based on similar test work. In their *Timber Frame Housing Design Guide*, TRADA also indicate values for the use of plasterboard clad partitions as racking panels. However, there are those who, for various reasons, believe that plasterboard should not be considered as a suitable sheathing agent. One of these reasons is that it is considered possible for the plasterboard to lose strength over a protracted period of time. In the opinion of the author, though this argument may have some merit for external walls, where internal walls are concerned, the risk of such deterioration would seem to be negligible. In modern standards of construction where there is an ever-increasing emphasis on moisture removal and high thermal insulation values, then it becomes increasingly difficult to substantiate deterioration of the material by age alone. It, therefore, seems sensible to allow a degree of usage of plasterboard clad partitions on internal wall panels.

It is difficult to be precise in regard to the extent of this degree of usage and until further information is available, it must be left largely to the experience of the individual designer. As a general guide, it is suggested that a range of 15–20% of the total racking resistance would not be unreasonable when considering the contribution which can be made by plasterboard clad internal partitions.

Example 11.3

To serve to illustrate the design of the racking resistance of an individual wall, let us examine the more common usage when applied to the construction of a

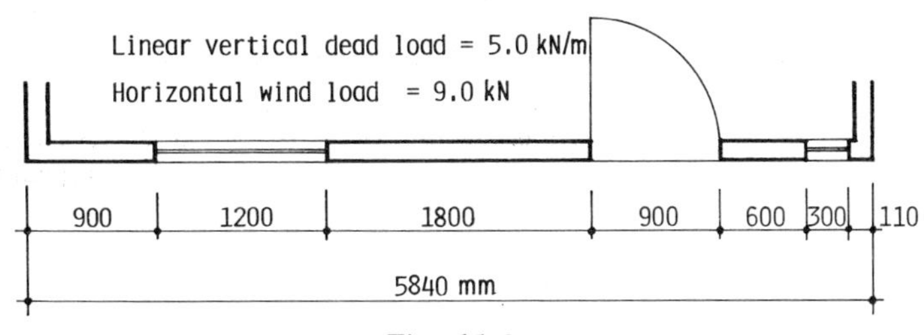

Fig. 11.9

timber framed domestic dwelling. Figure 11.9 gives the plan view of an external wall panel where both the vertically applied dead loads and horizontally applied wind loads have been pre-determined. It will be assumed that all openings will be ignored and all sections of the wall less than 600 mm in length make no contribution to the total resistance. Using the value given in Table 11.1 determine the total horizontal resistance of the wall. Comment on fixings.

Studs are at 600 mm c/c and so

$$\text{axial load per stud} = 5.0 \times 0.6 = 3.0\ \text{kN}$$

From Table 11.1 by interpolation

$$\text{permissible racking resistance} = 3.4 + \left(\frac{0.5}{2.5} \times 0.50\right)$$

$$= 3.5\ \text{kN/m}$$

Total effective linear run in wall = 0.90 + 1.80 + 0.60 = 3.3 m

Racking resistance = 3.5 × 3.3 = 11.55 kN > 9.0 kN applied.

Check rotational uplift: Assuming the overall panel heights to be 2.4 m and also assuming that the door opening effectively breaks the panel into two sections then the horizontal load may be apportioned in accordance with the relative stiffnesses of lengths L_1 and L_2 as indicated in Fig. 11.10.

Effective total length resisting rotation = $L_1 + L_2$ = 4.94 m.

$$\text{Relative stiffness of } L_1 = \frac{L_1^2}{H} = \frac{3.9^2}{2.4} = 6.34$$

$$L_2 = \frac{L_2^2}{H} = \frac{1.04^2}{2.4} = 0.45$$

$$= 6.79$$

$$\text{Horizontal load resisted by } L_1 = 9.0 \times \frac{6.34}{6.79} = 8.40\ \text{kN}$$

$$L_2 = 9.0 - 8.40 = 0.60\ \text{kN}$$

Considering lengths L_1 and L_2 separately, Fig. 11.11 indicates the proportional loads acting upon them:

Panel L_1:

Overturning moment = 8.40 × 2.4 = −20.16

$$\text{Resisting moment} = 5.0 \times \frac{3.9^2}{2} = +38.00$$

Residual moment +17.84 kN m

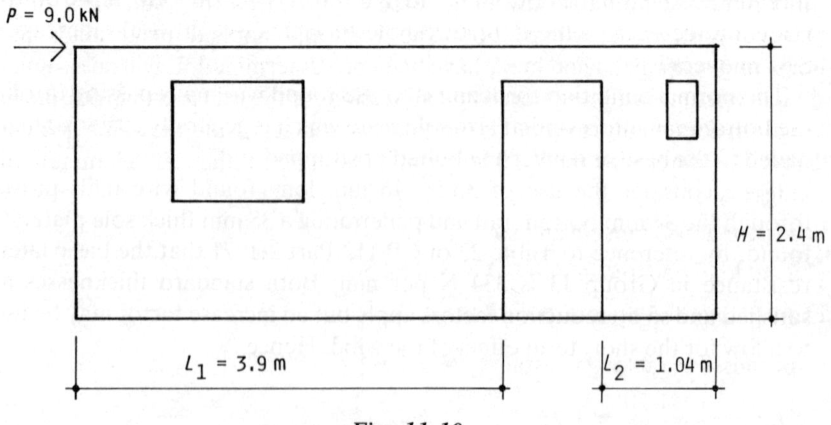

Fig. 11.10

Consequently uplift does not occur in this panel.

Panel L_2:

$$\text{Overturning moment} = 0.60 \times 2.4 = -1.44$$

$$\text{Resisting moment} = 5.0 \times \frac{1.04^2}{2} = +2.70$$

$$\text{Residual moment} \qquad +1.26 \text{ kN m}$$

Once again, uplift does not occur.

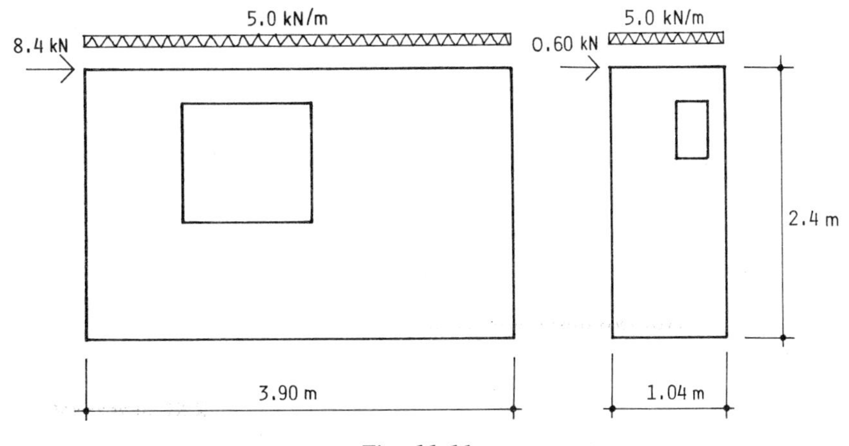

Fig. 11.11

Check horizontal fixings at base of panels: Although the wall is broken into two panels by virtue of the door opening, the total horizontal load may be spread linearly into the bottom rail length which is 4.94 m long. If the bottom rail continued across the door threshold then this length would increase to 5.84 m overall.

The normal condition for fixing is to use round wire nails passing through the bottom rail into a similar size sole plate which is generally either similarly nailed to the base or conversely bolted or strapped.

If we consider the use of 3.65×76 mm long round wire nails passing through the 38 mm bottom rail and penetrating a 38 mm thick sole plate, it is found, by reference to Table 22 of CP 112:Part 2:1971 that the basic lateral resistance in Group J3 is 334 N per nail. Both standard thicknesses are satisfied and so no reduction factors apply but an increase factor may be used to allow for the short term effect of the wind. Hence

$$K_{21} = \frac{1 + 1.50}{2} = 1.25$$

Giving

design resistance = $334 \times 1.25 = 417$ N per nail

$$\text{Total number of nails required} = \frac{9000}{417} = 22 \text{ No. min.}$$

11.2.3 Whole building effect

To conclude on this subject, it is worth considering the effect of the integration of the panels into the building as a whole. It is convenient to consider the panels as isolated elements resisting an assigned proportion of the lateral loads in accordance with their various stiffnesses, as described in the preceding example. There can be little doubt, however, that the addition of all finishes has a considerable stiffening effect and adds strength to the building. The quantification of this strength cannot currently be accurately defined and indeed it may never be possible to do so but it is possible to conclude that such approaches to the design as have been described are conservative.

11.3 Sacrificial timber to BS 5268:Part 4:1978, Section 4.1

11.3.1 Background

We have already touched upon the subject of fire in Section 5.6 when we examined the Building Regulations in regard to Part E of the regulations. However, the following few paragraphs will give a little more background to the subject before proceeding with design examples.

The predictability of the performance of both softwoods and hardwoods in fire is well established and its speed of burning, known as 'the charring rate', is largely related to the density of the timber. The more dense the timber, the slower will be the charring rate and, for the more commonly used softwoods, above a density of 420 kg/m^3, the depletion of the cross-section is 20 mm in 30 minutes of exposure, for each face. Below this density, the rate is taken to be 25 mm for the same period and most structural hardwoods may be taken as 15 mm in 30 minutes. The charring rate is considered to be linear and so a one hour period would require 40 mm, 50 mm and 30 mm, respectively.

Because timber is so predictable, it becomes relatively simple to design elements such as beams and columns and it is these two principal elements that we will eventually examine. In regard to composite construction, the task is not so easy and, consequently, we must largely rely upon the expedient of the fire test to provide the necessary ratings. Most proprietary methods of construction, where necessary, undergo fire test ratings and a test certificate is supplied to ensure that proper procedures have been adhered to.

The following is a list of British Standards which relate to various test procedures and are used to examine the various features which occur when a material is exposed to fire:

BS 476:Part 4:1970	Non-combustibility test for materials
BS 476:Part 5:1968	Ignitability test for materials
BS 476:Part 6:1968	Fire propagation test for materials
BS 476:Part 7:1971	Surface spread of flame tests for materials
BS 476:Part 8:1972	Test methods and criteria for the fire resistance of elements of building construction

As a further guide these publications may be further explained as follows:

(a) Non-combustibility

The term non-combustibility in the context of this test does not imply that it will not burn but the material is examined for its performance at a temperature of 750°C which is the temperature generally recognized as prevailing during the first stage of a fire.

(b) Ignitability

This test requires a small panel of the material to be exposed to a pilot flame for a short period of time and it is then classified on the basis of flame spread and sustained flaming. Timber and its derivatives are not easily ignitable because of the thicknesses generally used in building construction.

(c) Fire propagation

This test examines the rate of heat release of the material and hence its contribution to the prolongation of the fire. Lining materials are required under the Building Regulations to be Class 0 and timber falls well below this standard.

(d) Surface spread of flame

This is a measure of the rate and extent of travel of the front of a flame across a 900 mm long panel of the material to be tested. The test defines four categories with Class 1 the highest down to Class 4. Timber, plywood, particle board and hardboard are given Class 3 ratings provided that their density is at least 400 kg/m^3. Species below this density are graded Class 4. Upgrading to Class 1 is possible with the use of various specialist surface treatments or impregnation treatments.

(e) Fire resistance of elements

The fire resistance of an element is defined as its ability to continue to perform its function within a building despite being exposed to a fully developed fire. It is not necessarily related to the so-called fire resistance of the materials which form the element as other items such as fixings, thermal expansion, support details, etc., will influence its performance. Test methods are available for the more common elements of floors, walls, beams, columns, doors, etc.

Structural elements must be capable of sustaining their integrity for a specified period and fire barriers must resist the passage of fire or excessive heat. In all cases, the overall structural stability and fire containment must be achieved until the fire fighting authorities have brought the fire under control.

11.3.2 BS 5268:Part 4:1978, Section 4.1

This code enables the strength of flexural, tension and compression members to be assessed by calculation using stress modification factors which have been arrived at empirically and checked against the results of fire tests. The code may be used for assessing solid or glue laminated timber elements and their associated joint fixings. The examples which follow will serve to highlight the principles involved when designing beams and columns.

Example 11.4

A 100 mm × 300 mm deep beam is exposed to fire on three faces (Fig. 11.12) and a rating of a half hour fire resistance is required. Determine the strength of the beam, given that its effective span is 3.0 m and that it carries a uniformly distributed total load of 30 kN and the species is European Redwood, stress graded SS.

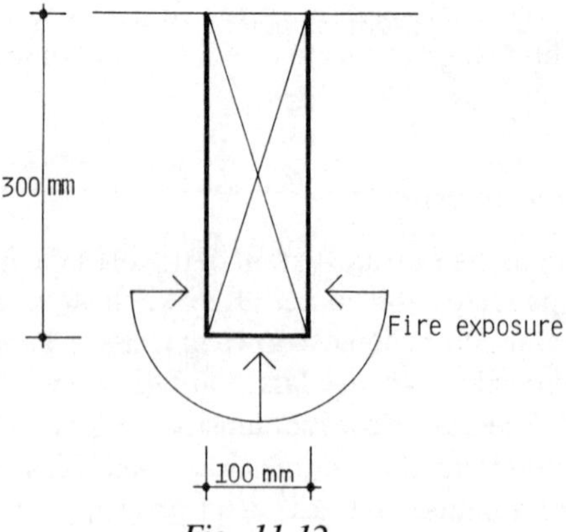

Fig. 11.12

Residual section and its properties: The code requires the residual section to be found by subtracting the notional amount of charring caused by the fire period. In this case, the species is Redwood and the period is ½ hour giving 20 mm on all faces as given in Table 1 of the code. This gives a residual section as indicated in Fig. 11.13 leading to the following properties: Clause 4.3 of the code permits the rounded arrises to be ignored where the least dimension is

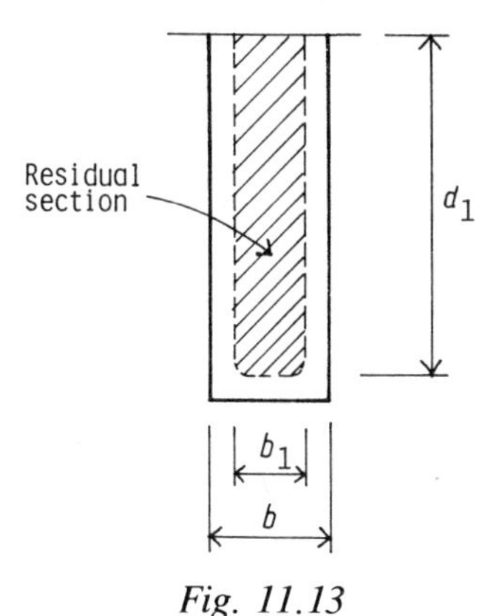

Fig. 11.13

not less than 50 mm, in this case the least dimension is 100 mm and so the rounding effect may be ignored:

$$I = \frac{b_1 d_1^3}{12} = \frac{6.0 \times 28.0^3}{12} = 10\,976\ \text{cm}^4$$

$$Z = \frac{b_1 d_1^2}{6} = \frac{6.0 \times 28.0^2}{6} = 784\ \text{cm}^3$$

$$A = b_1 d_1 = 6.0 \times 28.0 = 168\ \text{cm}^2$$

Moment capacity: The moment capacity

$$\bar{M} = Z \times 2.25 f_{\text{ppar}}$$

where

$$2.25 = \text{stress increase factor where } b > 70\ \text{mm}$$

$$f_{\text{ppar}} = 7.3\ \text{N/mm}^2\ \text{(Table 3.7)}$$

$$\bar{M} = 784 \times 2.25 \times \frac{7.3}{1000}$$

$$= 12.87\ \text{kN m}$$

Applied $$M = 30 \times \frac{3.0}{8} = 11.25\ \text{kN m} < \bar{M}$$

Shear capacity $\bar{V} = 2.25 v_{\text{ppar}} \times \dfrac{2A}{3}$

$$= 2.25 \times \frac{0.86}{10} \times 2 \times \frac{168}{3}$$

$$= 21.67 \text{ kN}$$

Applied $V = 30 \times \tfrac{1}{2} = 15 \text{ kN} < \bar{V}$

Deflection $d = \dfrac{5WL^3}{384EI}$ where $E = 5700 \text{ N/mm}^2$

$$d = \frac{5 \times 30\,000 \times 3000^3}{384 \times 5700 \times 10\,976 \times 10^4} = 16.85 \text{ mm}$$

To comply with Clause 5.1.1(b) the applied deflection must not exceed

$$\frac{L}{30} = \frac{3000}{30} = 100 \text{ mm} > 16.85$$

The conclusion of this investigation is that the cross-section contains sufficient sacrificial timber to satisfy the design requirement of a half hour fire resisting period.

Example 11.5

An isolated axially loaded column measuring 150 mm square and 2.4 m high carries a concentric load. It is GS graded Whitewood and required to withstand a fire for a period of one hour. Determine the maximum design load which may be carried.

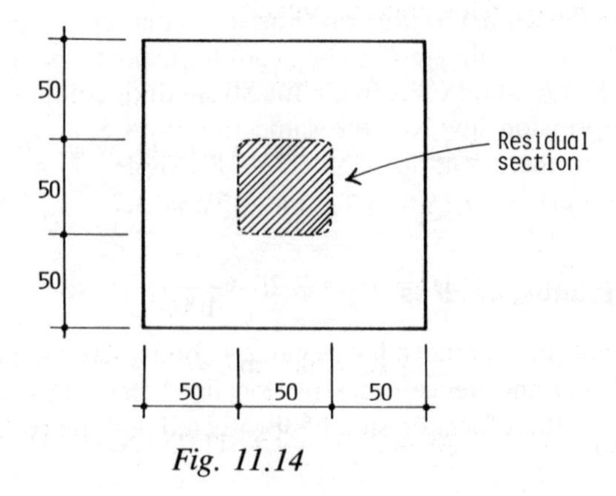

Fig. 11.14

Residual section and its properties: Where a fire is able to attack all four faces at once, the rate of charring given in Table 1 must be increased by a factor of 1.25. For the one hour period required, this would amount to $40 \times 1.25 = 50$ mm all round (Fig. 11.14). Once again, the rounded arrises may be ignored giving the following properties:

$$A_r = 25 \text{ cm}^2 \qquad I = 5 \times \frac{5^3}{12} = 52.0 \text{ cm}^4$$

$$r = \sqrt{\frac{52}{25}} = 1.44 \text{ cm}$$

It will be assumed that to comply with Clause 5.2.2(b) the effective column height = 2400 mm, giving

$$\frac{l}{r} = \frac{2400}{14.4} = 166.6 < 250 \text{ satisfactory.}$$

From Table 15 of CP 112 $K_{18} = 0.226$ (long term).
The load capacity of the column (P) may now be determined from the equation

$$\bar{P} = 2.0 \times K_{18} \times C_{\text{ppar}} \times A_r$$

where

$$2.0 = \text{stress increase factor (Clause 5.2.2(d))}$$

$$C_{\text{ppar}} = 5.6 \text{ N/mm}^2 \text{ (Table 3.7)}$$

$$\bar{P} = 2.0 \times 0.226 \times 5.6 \times \frac{25}{10} = 6.33 \text{ kN}$$

This example serves to illustrate the dramatic effect that large notional fire periods have on what otherwise would appear to be a substantial column section. The method of deriving the strength of columns subject to bending plus compression involves the same theory as Section 10.2 with the combination of the limitations given in Examples 11.4 (bending) and 11.5 (compression).

11.3.4 Jointing devices

Where joints are composed of metal, e.g. bolts, screws, nails, plates, etc., the path for heat and hence excessive localized charring is enhanced. All such joints must, therefore, be suitably protected and the code allows one of two methods:

(a) Embedding the connector to at least the same depth as that required for the sacrificial timber and suitably fire stopping all holes, etc. The plugging should normally be glued timber unless other specialist advice prevails.
(b) Covering the exposed fasteners with a suitable fire resisting material which gives the notional fire period required. Exposed nails, screws or staples may be used to fix this protecting material but special attention should be paid to the detail to ensure that the material remains in place for the required period of fire resistance.

From these two choices, it would seem prudent to use method (a) wherever possible because, although undoubtedly it would prove the more expensive, it is a more positive approach to solving the problem and gives a cleaner appearance.

12
Jointing fasteners

12.1 General

The types of jointing fasteners are numerous and varied ranging from the common round wire nail through to many proprietary timber connectors such as truss clips and framing anchors. Prices are wide ranging and competition is keen.

The choice by the designer of the particular connecting agent will depend upon the actual design problem to be solved; however, by far the commonest fastener is the round wire nail and so it is with this that we will start and from this progress through some of the other means that are easily available to the designer.

12.2 Round wire nails

There are many occasions in joining timbers together with round wire nails when little, if any, load is applied to the joint and, because of this, the nails' versatility and ease of usage is well recognized. Where loads do occur, they are light in magnitude and invariably apply shear to the nail. Tests have determined that the amount of shear which may be taken is controlled by the thicknesses of the pieces of timber to be joined together, length and diameter of the nail used and the specific gravity and moisture content of the wood. From these tests, it has been shown that there is no significant difference for loads applied parallel or perpendicular to the grain. Failure is caused by bending in the nail and crushing of the timber. Table 12.1 indicates some typical stock sizes and lengths for round wire nails.

12.2.1 Single shear

CP 112:Part 2, Clause 3.20.2.3 gives the permissible basic loads for round wire nails in single shear for various pointside and headside thicknesses and various groups (grouped species) of timber. The simple definition of pointside

Table 12.1 Round plain head nails

Length (mm)		*Thickness (mm)*	*Length (inches)*		*Gauge*
150	×	6.0	6	×	4
125	×	5.6	5	×	5
125	×	5.0	5	×	6
115	×	5.0	4½	×	6
100	×	5.0	4	×	6
100	×	4.5	4	×	7
100	×	4.0	4	×	8
90	×	4.0	3½	×	8
75	×	4.0	3	×	8
75	×	3.75	3	×	9
75	×	3.35	3	×	10
65	×	3.35	2½	×	10
65	×	3.0	2½	×	11
65	×	2.65	2½	×	12
60	×	3.35	2¼	×	10
60	×	3.0	2¼	×	11
60	×	2.65	2¼	×	12
50	×	3.35	2	×	10
50	×	3.0	2	×	11
50	×	2.65	2	×	12
50	×	2.36	2	×	13
45	×	2.65	1¾	×	12
45	×	2.36	1¾	×	13
40	×	2.65	1½	×	12
40	×	2.36	1½	×	13
40	×	2.0	1½	×	14
30	×	2.36	1¼	×	13
30	×	2.0	1¼	×	14
25	×	2.0	1	×	14
25	×	1.8	1	×	15
25	×	1.6	1	×	16
20	×	1.6	¾	×	16

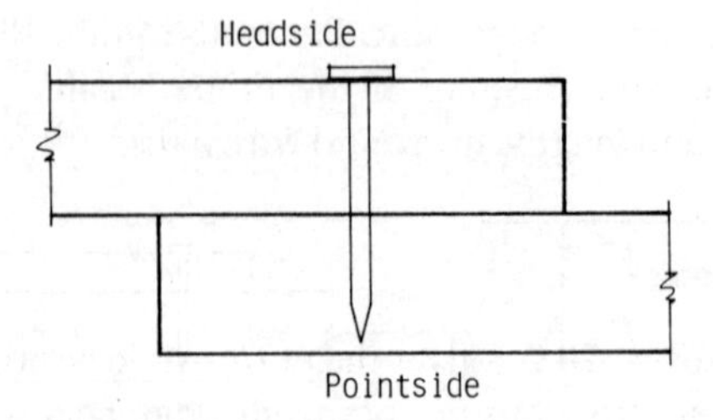

Fig. 12.1

Table 12.2 Dry basic lateral loads in single shear for round wire nails inserted at right-angles to the side grain

Nail		*Standard thickness of members*		*Basic lateral load per nail for timbers in group*			
Diameter (mm)	*(SWG)*	*Headside (mm)*	*Pointside (mm)*	*J1 (N)*	*J2 (N)*	*J3 (N)*	*J4 (N)*
2.64	12	19	25	245	200	178	133
2.95	11	22	29	311	245	222	178
3.25	10	25	32	378	311	267	222
3.66	9	29	38	489	378	334	267
4.06	8	32	44	623	489	400	334
4.47	7	38	51	712	578	489	400
4.88	6	44	57	890	667	578	489
5.38	5	51	67	1070	845	712	578
5.89	4	57	76	1250	979	845	667
6.40	3	64	89	1380	1110	934	756
7.94	—	83	108	1600	1290	1070	845
9.52	—	95	127	2050	1690	1380	1110

and headside is illustrated in Fig. 12.1 and the basic loads are reproduced in Table 12.2. The code does not permit an increase in these basic loads with an increase in thickness but, conversely, the basic load must be reduced if a thinner thickness is used and this reduction is calculated in proportion to the tabulated thickness.

An example of calculating basic nail joint resistance follows.

Example 12.1

The joint in Fig. 12.2 is subjected to a long term tensile load; determine its load capacity when using Group J3 timber.

From Table 12.2 the pointside penetration must be 38 mm and so this is fully satisfied but the headside of 25 mm is less than the 29 mm required, therefore, there is a proportional reduction in the basic lateral load as follows:

$$\text{basic load (Group J3)} = 334 \times \frac{25}{29} = 288 \text{ N long term per nail}$$

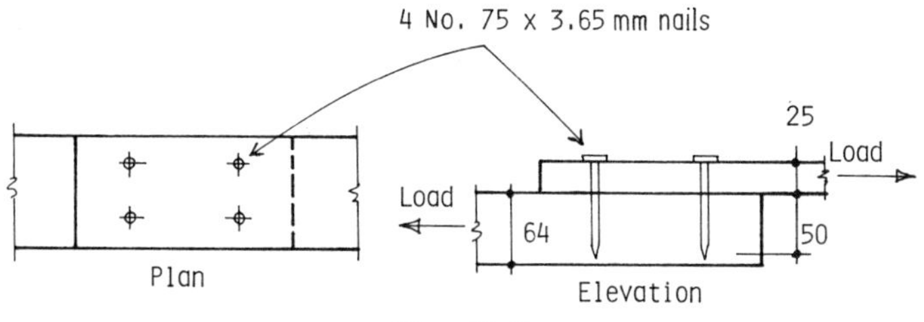

Fig. 12.2

therefore

$$\text{joint capacity} = 288 \times 4 = 1152\ \text{N}$$

Had the timber used been in the green condition then this capacity would have had to be reduced by multiplying by 0.7 to comply with Clause 3.20.4.

To avoid undue splitting, the spacing of nails in this joint must conform with Clause 3.20.6 and had the headside timber been replaced by a metal plate of sufficient strength, the basic load could be increased by 25% (×1.25). This allows for the increase in shear strength of the nail provided that it is driven through a tight fitting hole.

Where nails are subjected to term loading an increase factor K_{21} may be introduced where

$$K_{21} = \frac{1 + K_{12}}{2}$$

and K_{12} = the increase factor for term loading as described in Clause 3.12.1.1.

In Clause 3.20.2.1 the lateral load for nailing into the end grain is found by multiplying the lateral load through the side grain by 0.67.

Example 12.2

The loaded connection in Fig. 12.3 is caused by load from a roof construction; determine the capacity of the joint in Group J3 timber.

$$\begin{array}{l} \text{Headside capacity} \\ \text{(side grain)} \end{array} = \frac{\text{actual thickness}}{\text{tabulated thickness}} \times \text{basic J3} \times K_{21}\ (\text{med. term})$$

$$= \frac{25}{32} \times 400 \times \frac{1 + 1.25}{2} = 351\ \text{N}$$

$$\begin{array}{l} \text{Pointside capacity} \\ \text{(end grain)} \end{array} = \frac{\text{actual penetration}}{\text{tabulated penetration}} \times \text{basic J3} \times K_{21} \times 0.67$$

$$= 1.0 \times 400 \times \frac{1 + 1.25}{2} \times 0.67 = 301\ \text{N}$$

Therefore in this example the pointside capacity is the limiting factor in the joint's strength.

It should be noted that the actual penetration pointside was greater than the tabulated and so, because an increase is not allowed, unity is maintained.

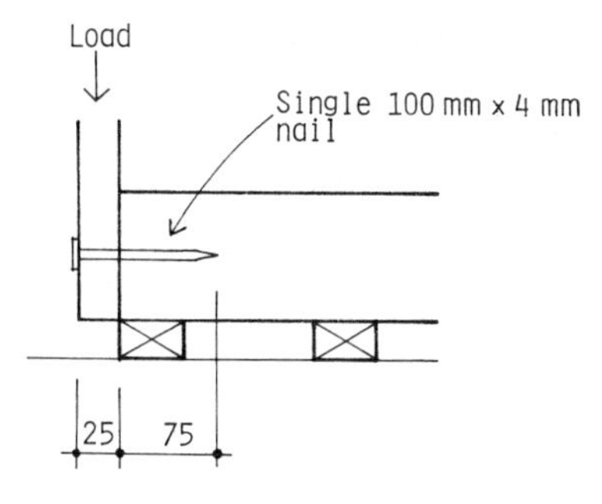

Fig. 12.3

12.2.2 Double shear

Where nails are in double shear Clause 3.20.2.2 stipulates that the basic loads should be multiplied by 0.9 times the number of shear planes. However, a further stipulation is that each member in a three member joint must have at least 0.7 times the thickness required for the pointside.

Example 12.3

The joint shown in Fig. 12.4 is subjecting the nail to double shear; find the long term capacity of the nail in Group J2 timber.

Checking the minimum supplied thickness against 0.7 of the tabulated thickness we have:

$$44 \times 0.7 = 30.8\ \text{mm} < 35\ \text{mm}$$

which is satisfactory.

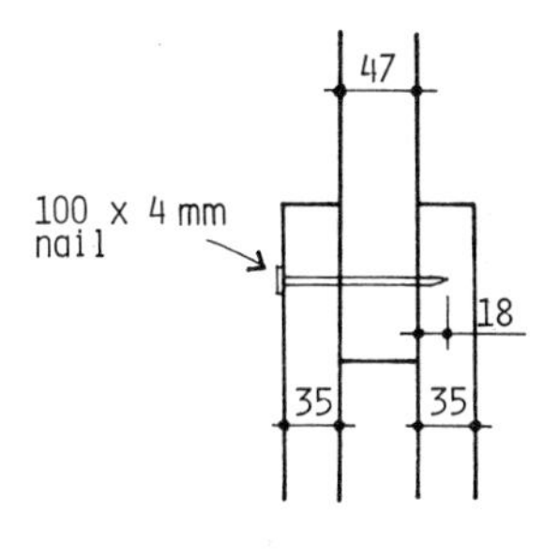

Fig. 12.4

In this joint the capacity will be limited by the pointside penetration of 18 mm.

$$\text{basic nail capacity} = \frac{18}{44} \times 489 = 200\ \text{N}$$

$$\text{joint capacity} = 200 \times 2 \text{ shear planes} \times 0.90$$

$$= 360\ \text{N}$$

Other factors, such as thickness, green timber, term loading and spacing of nails, discussed for single shear conditions, apply equally to double shear.

12.2.3 Withdrawal loads

Round wire nails are extremely weak in withdrawal and, consequently, they should not be relied upon to any great extent in joint design. Clause 3.20.2.1 gives basic withdrawal loads per millimetre of penetration for similar groups of species and these are reproduced in Table 12.3. These basic loads apply to both green and dry conditions. It is not permissible to take for loads in withdrawal when the nail is driven into the end grain of timber. To allow for changes in moisture content, i.e. swelling and shrinkage, the basic loads must be multiplied by 0.25 if such changes in moisture content are seen to exist.

Table 12.3 Basic resistance to withdrawal of round wire nails inserted at right-angles to the grain

Nail		*Basic resistance to withdrawal for timbers in group (expressed in N/mm of penetration)*			
Diameter (mm)	*SWG*	*J1*	*J2*	*J3*	*J4*
2.6	12	5.78	3.85	2.28	1.40
2.9	11	6.48	4.38	2.63	1.58
3.2	10	7.18	4.73	2.98	1.75
3.7	9	8.05	5.43	3.33	1.93
4.0	8	8.93	5.95	3.68	2.10
4.5	7	9.81	6.48	4.03	2.28
4.9	6	10.7	7.18	4.38	2.63
5.4	5	11.9	7.88	4.90	2.80
5.9	4	13.0	8.58	5.25	3.15
6.4	3	14.2	9.46	5.78	3.50
7.9	—	17.5	11.7	7.18	4.20
9.5	—	21.0	14.0	8.58	5.08

Example 12.4

The cladding rails shown in Fig. 12.5 are subjected to a wind suctional force; determine the resistance to withdrawal of the fixing nail in Group J3 timber.

From Table 12.3 the basic resistance to withdrawal = 3.33 N/mm. The term increase factor

$$K_{21} = \frac{1 + 1.50}{2} = 1.25$$

and so

$$\text{total resistance} = 3.33 \times 1.25 \times 50 \text{ mm penetration}$$
$$= 208 \text{ N}$$

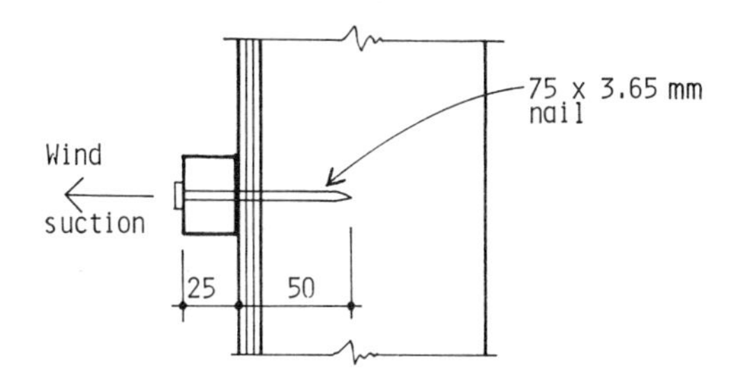

Fig. 12.5

If this joint were to be unprotected, then this resistance would be reduced to $208 \times 0.25 = 52$ N which is a significant reduction.

12.3 Improved nails

For the majority of usages improved nails may be considered as being either square twisted in form or annular ringed. Square twisted nails may be used in accordance with Clause 3.20.2.4 to improve the shear capacity of a joint by multiplying the basic loads by 1.25.

Annular ringed nails are primarily used where resistance to withdrawal is of prime consideration. An example of this would be in fixing certain types of sheathing subject to cyclic loading such as flooring chipboard.

Clause 3.20.3.3 permits the basic withdrawal loads to be increased by 25% when these nails are used and also there is no requirement to reduce the load for moisture changes. Some standard stock sizes are given in Table 12.4 for annular ringed and helical threaded nails.

Now, if we examine once again Fig. 12.5 and introduce ring shanked nails, then the joint resistance, even for green conditions, would be

$$\text{total withdrawal resistance} = 208 \times 1.25 = 260 \text{ N}$$

Therefore the value of such nails in withdrawal is clearly demonstrated.

Table 12.4
(a) Annular ringed nails

Length (mm)		*Thickness (mm)*	*Length (inches)*		*Gauge*
100	×	5.0	4	×	6
75	×	3.75	3	×	9
65	×	3.35	2½	×	10
60	×	3.35	2¼	×	10
50	×	3.35	2	×	10
50	×	3.0	2	×	11
50	×	2.65	2	×	12
45	×	2.65	1¾	×	12
40	×	2.65	1½	×	12
40	×	2.36	1½	×	13
30	×	2.36	1¼	×	13
25	×	2.0	1	×	14
20	×	2.0	¾	×	14

(b) Helical threaded nails

Length (mm)		*Thickness (mm)*	*Length (inches)*		*Gauge*
75	×	3.75	3	×	9
65	×	3.75	2½	×	9
65	×	3.35	2½	×	10
60	×	3.0	2¼	×	11
60	×	2.65	2¼	×	12
50	×	3.0	2	×	11
40	×	2.65	1½	×	12

12.4 Wood screws

There are many occasions when for various reasons wood screws are used in preference to nails but it should be clearly understood that the manner of fixing, unlike the nail, is not by using a hammer. Screws must always be properly threaded into the timber by means of a screwdriver and the basic values for lateral and withdrawal loads, given in the code, are based upon this assumption. The use of a hammer to drive screws into wood, will negate these design values.

12.4.1 Design features

The principal design features for the use of wood screws are similar to those expanded upon for round wire nails, with the exception that Tables 12.5 and 12.6 give the permissible basic lateral and withdrawal loads respectively, as defined in CP 112:Part 2. In addition, the code allows a minimum penetration equal to but not less than 0.6 times the standard tabulated value.

Table 12.5 Dry basic lateral loads for wood screws inserted at right angles to the grain

Screw		*Standard penetration*	*Basic lateral load for timbers in group*			
No.	*Diameter (mm)*	*(mm)*	*J1 (N)*	*J2 (N)*	*J3 (N)*	*J4 (N)*
4	2.7	19	205	169	138	107
5	3.1	22	262	218	178	138
6	3.4	25	329	271	222	173
7	3.8	27	400	329	267	209
8	4.2	29	476	391	320	249
9	4.5	32	560	463	378	294
10	4.9	35	654	538	440	342
11	5.2	38	752	623	507	396
12	5.6	41	858	707	578	449
14	6.3	44	1090	903	738	574
16	7.0	51	1350	1120	912	712
18	7.7	54	1640	1360	1110	963
20	8.4	60	1960	1610	1320	1030

Table 12.6 Dry basic resistance to withdrawal of wood screws inserted at right angles to the grain

Screw		*Basic resistance to withdrawal for timbers in group (expressed in N/mm of penetration of threaded part of screws)*			
No.	*Diameter (mm)*	*J1*	*J2*	*J3*	*J4*
4	2.7	16.1	11.6	7.88	5.08
5	3.1	18.2	13.1	8.93	5.78
6	3.4	20.3	14.7	9.98	6.48
7	3.8	22.4	16.1	10.8	7.18
8	4.2	24.5	17.7	11.9	7.88
9	4.5	26.6	19.3	13.0	8.58
10	4.9	28.7	20.7	14.0	9.11
11	5.2	30.6	22.2	15.1	9.81
12	5.6	32.7	23.6	16.1	10.5
14	6.3	36.9	26.8	18.0	11.9
16	7.0	41.1	29.8	20.1	13.1
18	7.7	45.4	32.7	22.2	14.5
20	8.4	49.6	35.7	24.3	15.9

12.4.2 Spacing

The code of practice refers to Clauses 3.21.1 and 5.4.2 for pre-drilling of holes to receive screws, with the spacing the same as for nails.

In Clause 5.4.2 it states that the shank of the screw should have a lead hole of equal diameter while the threaded portion should be driven into a hole not exceeding 0.9 times the root of the thread nearest to the shank. Manufacturers of screws will normally give their recommendations for this

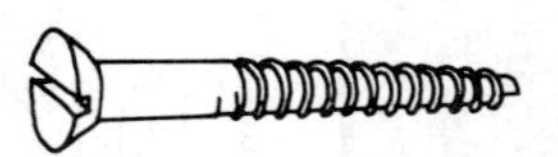

Table 12.7 Slotted steel wood screws, countersunk

Screw		*Lengths available (mm) (√ preferred, × special order)*																					
Gauge	*Diameter (mm)*	6.4	9.5	12.7	15.9	19.0	22.2	25.4	28.6	31.8	38.1	44.5	50.8	57.2	63.5	69.9	76.2	82.5	88.9	101.6	114.3	127	152.4
0	1.52	√																					
1	1.78	√	√																				
2	2.08	√	√	√																			
3	2.39		√	√	√	×		×															
4	2.74	×	√	√	√	√	×	√		√	√												
5	3.10			√	√	√		√		√													
6	3.45		×	√	√	√	×	√	×	√	√	×	√	√	√		√						
7	3.81			√	√	√	×	√		√	√	√	√										
8	4.17		×	√	√	√	×	√	×	√	√	√	√	×	√	×	√		√	√			
9	4.52					√		√		√	√	√	√		×								
10	4.88			×	×	√		√		√	√	√	√	√	√	×	√	√	√	√	√	√	
12	5.59					×		√		√	√	√	√	√	√	×	√	×	√	√	√	√	√
14	6.30							×		√	√	√	√	√	√	×	√		√	√	√	√	√
16	7.01										√	√	√		√		√		√	√		√	√
18	7.72												√				√			√			
20	8.43												√				√			√	√		

procedure but it has to be said that on the site, it is unusual for this procedure to be adopted. Normally, only one hole is drilled and this may not always be the 0.9 times expected. However, where factory controlled conditions prevail, the full procedure can be obtained because of closer quality control.

12.4.3 Screw types

There are several types of screws available, but the most common is the slotted countersunk head which is generally driven into a countersunk hole to finish flush with the surface of the timber. Others available are slotted round heads, good for fixing metal plates without the need to countersink and other proprietary screws which possess straight shanks and are less prone to splitting the timber than ordinary screws. Coach screws are another type which are suitable for carrying large loads and are fixed by use of a spanner applied to the square head. The range of stock sizes available for the more commonly used countersunk and round head are shown in Tables 12.7 and 12.8 with coach screws given in Table 12.9.

Table 12.8 Slotted steel wood screws, round

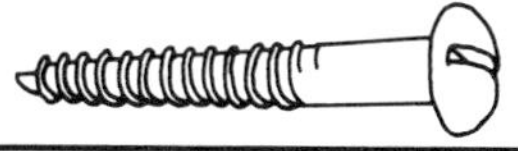

Screw		*Lengths available (mm) (√ preferred, × special order)*											
Gauge	*Diameter (mm)*	*6.4*	*9.5*	*12.7*	*15.9*	*19.0*	*25.4*	*31.8*	*38.1*	*44.5*	*50.8*	*63.5*	*76.2*
0	1.52	√											
2	2.08	√	√										
3	2.39		√	√									
4	2.74		√	√	√	√	√						
5	3.10			√	√	√							
6	3.45		√	√	√	√	√	√	√		×		
7	3.81				√	√	√	√					
8	4.17			√	√	√	√	√	√	√	√	×	
10	4.88				√	√	√	√	√	√	√	√	×
12	5.59					×	√	×	√	×	√	×	×
14	6.30								√		√		

Table 12.9 Steel coach screws, square heads

Screw diameter (mm)	*Lengths available (mm) (√ stock sizes)*												
	25.4	*31.8*	*38.1*	*44.5*	*50.8*	*57.2*	*63.5*	*76.2*	*88.9*	*101.6*	*114.3*	*127*	*152.4*
6.4	√	√	√	√	√	√	√	√	√	√		√	
7.9	√	√	√	√	√	√	√	√	√	√	√	√	√
9.5	√	√	√	√	√	√	√	√	√	√	√	√	√
12.7			√		√		√	√	√	√	√	√	√

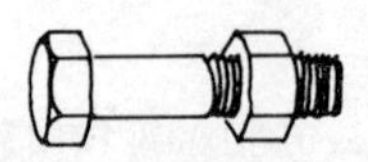

Table 12.10 Black hexagon bolts

Length under head (in)	*Diameter of bolt (in) (√ preferred items, × non-preferred items)*											
	1/4	*5/16*	*3/8*	*7/16*	*1/2*	*5/8*	*3/4*	*7/8*	*1*	*9/8*	*5/4*	*3/2*
1	√	√	√	√	√	√						
1 1/4	√	√	√	√	√	√						
1 1/2	√	√	√	√	√	√						
1 3/4	√	√	√	√	×	×						
2	√	√	√	√	√	√						
2 1/4	×	×	√	√	×	×						
2 1/2	√	√	√	√	√	√	√					
2 3/4	×	×	×	×	×	×	×					
3	√	√	√	√	√	√	√	√	√			
3 1/4	×	×	×	×	×	×	√	√				
3 1/2	√	√	√	√	√	√	√	√	√	√		
3 3/4	×	×	×	×	√	√	√	√				
4	√	√	√	√	√	√	√	√	√	√	√	
4 1/4	×	×	√	×	×	×	√	√				
4 1/2	√	√	√	√	√	√	√	√	√	√	√	
4 3/4	×	×	×	×	×	×	×	√	√			
5	√	√	√	√	√	√	√	√	√	√	√	
5 1/4	√	√	√	×	×	×	×	×	×			
5 1/2	√	√	√	√	√	√	√	√	√	√	×	
5 3/4	×	×	×	×	×	×	×	×	×			
6	√	√	√	√	√	√	√	√	√	√	√	√
6 1/4					×	×	×	×	×			
6 1/2					√	√	√	√	√		×	√
6 3/4					×	×	×	×	√			
7					√	√	√	√	√	√	√	√
7 1/4					×	×	×	×	×			
7 1/2					√	√	√	√	√	√	√	
7 3/4					×	×	√	×	×			
8					√	√	√	√	√	√	√	√
8 1/2					×	×	√	√	√	√		
9					√	√	√	√	√	√	√	√
9 1/2					×	×	√	√	√	√	×	
10					√	√	√	√	√	√	√	√
10 1/2					×	×	√	√	√	√	√	
11					√	√	√	√	√	√	√	√
11 1/2					×	×	√	√	√	√	√	
12					√	√	√	√	√	√	√	√

Note: Imperial sizes are now almost completely replaced by the metric sizes given in Table 12.11.

12.5 Bolts

All bolts specified in timber are assumed to be black bolts in accordance with BS916. The two sizes available are Imperial (inches) and ISO (metric) giving a wide choice to the designer. These are reproduced in Tables 12.10 and 12.11.

Where bolts are to be exposed either to the weather or as a feature, they

Table 12.11 Hex round hex, BS 4190/metric coarse standard thread length, grade 4.6 (√ preferred items, × non-preferred items)

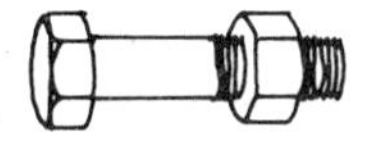

Length	*M6*	*M8*	*M10*	*M12*	*M16*	*M20*	*M24*	*M30*
20	√	×	×					
25	√	√	√	√				
30	√	√	√	√				
35	√	√	√	√				
40	√	√	√	√				
45	√	√	√	√				
50	√	√	√	√	√			
55	×	×	√	√	√			
60	√	√	√	√	√	√		
65	×	×	√	√	√	√		
70	√	√	√	√	√	√	√	
75	×	×	√	√	√	√	√	
80	√	√	√	√	√	√	√	
85	×	×	×	×	×	×	×	
90	√	√	√	√	√	√	√	
100	√	√	√	√	√	√	√	√
110	×	×	×	√	√	√	√	×
120	×	√	√	√	√	√	√	×
130	×	×	×	√	√	√	×	√
140	×	×	√	√	√	√	√	×
150	×	×	×	√	√	√	×	√
160				√	√	√	√	×
170				√	×	×	×	×
180				√	√	√	√	×
190				×	×	×	×	×
200				√	√	√	√	√
220				√	√	√	√	×
240				×	×	×	×	
260				√	√	√	√	×
280				×	×	×	×	
300				√	√	√	√	×

should be protected and this is best achieved by electrogalvanizing to give the correct finish and ease of fitting the nut to the bolt.

12.5.1 Basic single shear loads

For screws and nails, previously described, the direction of the load to the grain is not important but, with bolts where significant bearing areas apply, grain direction is important. Therefore, the dry basic loads given in the code and reproduced in Table 12.12 relate to the various groups with loads parallel and perpendicular to the grain.

From tests carried out on bolted joints, it has been established that depending on the ratio L/d the capacity of the bolt is governed either by

Table 12.12 Dry basic loads on one bolt in a two member joint

Thickness of thinner member	*Diameter of bolt*		*Basic load for one bolt in timbers of group*							
			J1		*J2*		*J3*		*J4*	
			Parallel to grain	*Perpendicular to grain*	*Parallel to grain*	*Perpendicular to grain*	*Parallel to grain*	*Perpendicular to grain*	*Parallel to grain*	*Perpendicular to grain*
(mm)	*(mm)*	*(inch)*	*(kN)*	*(kN)*	*(kN)*	*(kN)*	*(kN)*	*(kN)*	*(kN)*	*(kN)*
16	9.5	3/8	1.73	1.30	1.20	0.734	1.02	0.552	0.672	0.418
	12.7	1/2	2.33	1.41	1.61	0.818	1.35	0.614	0.903	0.467
	15.9	5/8	2.85	1.60	2.01	0.925	1.72	0.689	1.13	0.534
	19.0	3/4	3.38	1.69	2.35	1.05	2.02	0.778	1.24	0.578
	22.2	7/8	4.36	1.95	2.80	1.15	2.38	0.845	1.60	0.667
	25.4	1	4.54	2.17	3.16	1.23	2.67	0.925	1.80	0.712
19	9.5	3/8	1.85	1.48	1.30	0.854	1.17	0.632	0.801	0.489
	12.7	1/2	2.65	1.69	1.85	0.979	1.58	0.729	1.08	0.556
	15.9	5/8	3.33	1.91	2.34	1.11	2.00	0.818	1.33	0.632
	19.0	3/4	4.00	2.00	2.78	1.25	2.40	0.925	1.52	0.703
	22.2	7/8	4.89	2.31	3.29	1.38	2.80	1.00	1.87	0.783
	25.4	1	5.41	2.54	3.74	1.49	3.20	1.09	2.14	0.858
22	9.5	3/8	1.96	2.13	1.38	0.987	1.28	0.729	0.916	0.560
	12.7	1/2	3.00	1.96	2.08	1.12	1.82	0.845	1.25	0.645
	15.9	5/8	3.82	2.20	2.67	1.28	2.32	0.943	1.55	0.734
	19.0	3/4	4.63	2.34	3.25	1.43	2.79	1.06	1.79	0.810
	22.2	7/8	5.54	2.69	3.78	1.58	3.25	1.16	2.18	0.907
	25.4	1	6.18	2.96	4.34	1.71	3.71	1.27	2.49	0.987
25	9.5	3/8	2.02	1.89	1.41	1.12	1.35	0.827	0.987	0.641
	12.7	1/2	3.27	2.22	2.30	1.29	2.05	0.961	1.41	0.734
	15.9	5/8	4.31	2.49	3.02	1.45	2.62	1.08	1.74	0.836
	19.0	3/4	5.21	2.78	3.69	1.63	3.16	1.21	2.09	0.925
	22.2	7/8	6.23	3.05	4.29	1.79	3.69	1.32	2.49	1.03
	25.4	1	7.05	3.35	4.92	1.96	4.23	1.45	2.80	1.11
29	9.5	3/8	2.02	2.11	1.41	1.29	1.36	0.965	1.02	0.738
	12.7	1/2	3.50	2.57	2.46	1.49	2.29	1.11	1.59	0.854
	15.9	5/8	4.86	2.89	3.37	1.69	2.98	1.25	2.02	0.970
	19.0	3/4	6.00	3.24	4.18	1.89	3.63	1.41	2.44	1.07
	22.2	7/8	7.12	3.54	4.96	2.09	4.27	1.53	2.87	1.19
	25.4	1	8.14	3.89	5.66	2.27	4.89	1.68	3.27	1.30
32	9.5	3/8	2.02	2.20	1.41	1.41	1.36	1.07	1.03	0.818
	12.7	1/2	3.57	2.82	2.50	1.65	2.36	1.23	1.71	0.943
	15.9	5/8	5.14	3.18	3.60	1.87	3.38	1.38	2.22	1.07
	19.0	3/4	6.58	3.56	4.59	2.08	4.00	1.55	2.69	1.18
	22.2	7/8	7.76	3.89	5.45	2.30	4.69	1.69	3.15	1.31
	25.4	1	8.94	4.29	6.23	2.49	5.38	1.86	3.58	1.42
33	9.5	3/8	2.02	2.21	1.41	1.43	1.36	1.09	1.03	0.836
	12.7	1/2	3.58	2.89	2.50	1.69	2.38	1.26	1.75	0.979
	15.9	5/8	5.25	3.27	3.67	1.92	3.36	1.42	2.28	1.10
	19.0	3/4	6.75	3.68	4.71	2.14	4.11	1.59	2.78	1.22
	22.2	7/8	7.96	4.03	5.63	2.38	4.85	1.75	3.25	1.35
	25.4	1	9.21	4.45	6.45	2.58	5.56	1.91	3.74	1.47

Table 12.12 Dry basic loads on one bolt in a two member joint (*cont.*)

Thickness of thinner member	*Diameter of bolt*		*Basic load for one bolt in timbers of group*							
			J1		*J2*		*J3*		*J4*	
			Parallel to grain	*Perpendicular to grain*	*Parallel to grain*	*Perpendicular to grain*	*Parallel to grain*	*Perpendicular to grain*	*Parallel to grain*	*Perpendicular to grain*
(mm)	*(mm)*	*(inch)*	*(kN)*	*(kN)*	*(kN)*	*(kN)*	*(kN)*	*(kN)*	*(kN)*	*(kN)*
35	9.5	3/8	2.02	2.22	1.41	1.48	1.36	1.16	1.03	0.881
	12.7	1/2	3.58	3.05	2.50	1.81	2.40	1.34	1.80	1.03
	15.9	5/8	5.37	3.47	3.77	2.04	3.48	1.51	2.40	1.16
	19.0	3/4	7.05	3.89	4.93	2.27	4.34	1.69	2.94	1.29
	22.2	7/8	8.43	4.29	5.94	2.51	5.16	1.85	3.42	1.43
	25.4	1	9.71	4.71	6.82	2.73	5.87	2.04	3.94	1.56
36	9.5	3/8	2.02	2.21	1.41	1.49	1.36	1.18	1.03	0.903
	12.7	1/2	3.58	3.11	2.50	1.85	2.42	1.38	1.81	1.06
	15.9	5/8	5.43	3.58	3.80	2.10	3.54	1.56	2.45	1.20
	19.0	3/4	7.21	4.00	5.02	2.34	4.45	1.74	3.02	1.33
	22.2	7/8	8.63	4.40	6.09	2.58	5.29	1.91	3.54	1.47
	25.4	1	9.94	4.83	6.98	2.80	6.03	2.09	4.03	1.60
37	9.5	3/8	2.02	2.21	1.41	1.50	1.36	1.20	1.03	0.921
	12.7	1/2	3.58	3.18	2.50	1.90	2.42	1.42	1.81	1.09
	15.9	5/8	5.47	3.67	3.83	2.16	3.58	1.60	2.50	1.23
	19.0	3/4	7.32	4.12	5.11	2.40	4.55	1.79	3.09	1.37
	22.2	7/8	8.90	4.54	6.27	2.67	5.45	1.97	3.65	1.51
	25.4	1	10.2	4.98	7.16	2.89	6.18	2.16	4.16	1.65
38	9.5	3/8	2.02	2.19	1.41	1.50	1.36	1.22	1.03	0.934
	12.7	1/2	3.58	3.23	2.50	1.94	2.43	1.45	1.81	1.12
	15.9	5/8	5.52	3.77	3.86	2.22	3.62	1.65	2.57	1.27
	19.0	3/4	7.41	4.23	5.18	2.46	4.65	1.84	3.16	1.41
	22.2	7/8	9.12	4.65	6.38	2.73	5.58	2.02	3.74	1.56
	25.4	1	10.5	5.08	7.36	2.96	6.34	2.20	4.29	1.69
40	9.5	3/8	2.02	2.17	1.41	1.50	1.36	1.24	1.03	0.952
	12.7	1/2	3.58	3.33	2.50	2.05	2.43	1.53	1.81	1.18
	15.9	5/8	5.56	3.97	3.89	2.34	3.68	1.73	2.66	1.33
	19.0	3/4	7.58	4.45	5.29	2.58	4.85	1.93	3.32	1.48
	22.2	7/8	9.43	4.89	6.63	2.87	5.87	2.12	3.93	1.64
	25.4	1	11.0	5.36	7.70	3.11	6.69	2.32	4.49	1.77
41	9.5	3/8	2.02	2.15	1.41	1.49	1.36	1.24	1.03	0.956
	12.7	1/2	3.58	3.36	2.50	2.09	2.43	1.57	1.81	1.20
	15.9	5/8	5.57	4.07	3.90	2.39	3.70	1.78	2.69	1.37
	19.0	3/4	7.65	4.57	5.35	2.65	4.94	1.98	3.39	1.51
	22.2	7/8	9.61	5.03	6.74	2.94	6.00	2.18	4.05	1.68
	25.4	1	11.3	5.52	7.92	3.20	6.85	2.39	4.60	1.82
44	9.5	3/8	2.02	2.10	1.41	1.47	1.36	1.23	1.03	0.852
	12.7	1/2	3.58	3.40	2.50	2.21	2.43	1.68	1.81	1.25
	15.9	5/8	5.58	4.35	3.91	2.57	3.77	1.91	2.80	1.47
	19.0	3/4	7.84	4.89	5.47	2.85	5.12	2.13	3.62	1.63
	22.2	7/8	10.1	5.38	7.01	3.16	6.41	2.34	4.31	1.80
	25.4	1	12.0	5.87	8.38	3.42	7.34	2.56	4.94	1.95

Table 12.12 Dry basic loads on one bolt in a two member joint (*cont.*)

Thickness of thinner member	*Diameter of bolt*		*Basic load for one bolt in timbers of group*							
			J1		*J2*		*J3*		*J4*	
			Parallel to grain	*Perpendicular to grain*	*Parallel to grain*	*Perpendicular to grain*	*Parallel to grain*	*Perpendicular to grain*	*Parallel to grain*	*Perpendicular to grain*
(mm)	*(mm)*	*(inch)*	*(kN)*	*(kN)*	*(kN)*	*(kN)*	*(kN)*	*(kN)*	*(kN)*	*(kN)*
47	9.5	3/8	2.02	2.05	1.41	1.45	1.36	1.21	1.03	0.934
	12.7	1/2	3.58	3.40	2.50	2.28	2.43	1.78	1.81	1.37
	15.9	5/8	5.58	4.56	3.92	2.74	3.79	2.04	2.85	1.56
	19.0	3/4	7.96	5.22	5.58	3.03	5.27	2.27	3.79	1.78
	22.2	7/8	10.4	5.77	7.25	3.37	6.67	2.50	4.60	1.92
	25.4	1	12.6	6.29	8.76	3.67	7.85	2.74	5.29	2.08
50	9.5	3/8	2.02	1.99	1.41	1.41	1.36	1.19	1.03	0.916
	12.7	1/2	3.58	3.37	2.50	2.30	2.43	1.85	1.81	1.43
	15.9	5/8	5.58	4.70	3.92	2.92	3.79	2.17	2.87	1.66
	19.0	3/4	8.06	5.54	5.65	3.23	5.38	2.41	3.93	1.85
	22.2	7/8	10.6	6.12	7.43	3.57	6.92	2.65	4.85	2.04
	25.4	1	13.1	6.67	9.19	3.90	8.27	2.90	5.60	2.20
60	12.7	1/2	3.58	3.18	2.50	2.26	2.43	1.87	1.81	1.45
	15.9	5/8	5.58	4.80	3.92	3.23	3.79	2.54	2.87	1.96
	19.0	3/4	8.10	6.25	5.65	3.87	5.47	2.89	4.07	2.22
	22.2	7/8	11.0	7.23	7.72	4.29	7.38	3.19	5.45	2.45
	25.4	1	14.1	8.01	9.85	4.67	9.23	3.48	6.49	2.66
	28.6	1 1/8	17.1	8.76	12.0	5.09	10.8	3.78	7.47	2.94
63	12.7	1/2	3.58	3.12	2.50	2.20	2.43	1.85	1.81	1.43
	15.9	5/8	5.58	4.76	3.92	3.26	3.79	2.62	2.87	2.01
	19.0	3/4	8.10	6.35	5.65	4.05	5.52	3.02	4.07	2.33
	22.2	7/8	11.0	7.56	7.72	4.51	7.43	3.34	5.54	2.56
	25.4	1	14.2	8.41	9.96	4.92	9.39	3.65	6.74	2.79
	28.6	1 1/8	17.5	9.16	12.3	5.34	11.3	3.98	7.78	3.07
72	15.9	5/8	5.58	4.59	3.92	3.20	3.79	2.67	2.87	2.05
	19.0	3/4	8.10	6.46	5.65	4.33	5.52	3.44	4.07	2.62
	22.2	7/8	11.0	8.14	7.72	5.08	7.43	3.83	5.58	2.94
	25.4	1	14.3	9.43	10.0	5.60	9.70	4.16	7.23	3.19
	28.6	1 1/8	18.1	9.56	12.6	6.12	12.0	4.54	8.58	3.51
75	15.9	5/8	5.58	4.53	3.92	3.16	3.79	2.66	2.87	2.05
	19.0	3/4	8.10	6.38	5.65	4.36	5.52	3.51	4.07	2.69
	22.2	7/8	11.0	8.18	7.72	5.25	7.43	3.98	5.60	3.05
	25.4	1	14.4	9.72	10.0	5.85	9.72	4.34	7.29	3.33
	28.6	1 1/8	18.2	10.8	12.7	6.38	12.1	4.74	8.81	3.65
97	15.9	5/8	5.58	4.00	3.92	2.86	3.79	2.50	2.87	1.93
	19.0	3/4	8.10	5.87	5.65	4.11	5.52	3.47	4.07	2.68
	22.2	7/8	11.0	8.01	7.72	5.54	7.43	4.58	5.60	3.53
	25.4	1	14.4	10.3	10.0	6.92	9.72	5.44	7.34	4.20
	28.6	1 1/8	18.2	12.6	12.7	8.10	12.3	6.09	9.30	4.69
	31.8	1 1/4	22.4	14.4	15.7	8.83	15.2	6.56	11.4	5.05

Table 12.12 Dry basic loads on one bolt in a two member joint (*cont.*)

Thickness of thinner member	*Diameter of bolt*		*Basic load for one bolt in timbers of group*							
			J1		*J2*		*J3*		*J4*	
			Parallel to grain	*Perpendicular to grain*	*Parallel to grain*	*Perpendicular to grain*	*Parallel to grain*	*Perpendicular to grain*	*Parallel to grain*	*Perpendicular to grain*
(mm)	*(mm)*	*(inch)*	*(kN)*	*(kN)*	*(kN)*	*(kN)*	*(kN)*	*(kN)*	*(kN)*	*(kN)*
100	15.9	5/8	5.58	3.92	3.92	2.81	3.79	2.49	2.87	1.91
	19.0	3/4	8.10	5.72	5.65	4.07	5.52	3.42	4.07	2.66
	22.2	7/8	11.0	7.94	7.72	5.49	7.43	4.59	5.60	3.53
	26.4	1	14.4	10.2	10.0	6.94	9.72	5.56	7.34	4.25
	28.6	1 1/8	18.2	12.7	12.7	8.23	12.3	6.27	9.30	4.83
	31.8	1 1/4	22.4	14.7	15.7	9.10	15.2	6.78	11.4	5.23
145	25.4	1	14.4	8.87	10.0	6.34	9.72	5.40	7.34	4.17
	28.6	1 1/8	18.2	11.5	12.7	8.10	12.3	6.81	9.30	5.25
	31.8	1 1/4	22.4	14.4	15.7	9.96	15.2	8.34	11.4	5.47
150	25.4	1	14.4	8.72	10.0	6.23	9.72	5.38	7.34	4.13
	28.6	1 1/8	18.2	11.3	12.7	7.98	12.3	6.76	9.30	5.16
	31.8	1 1/4	22.4	14.2	15.7	9.87	15.2	8.27	11.4	5.29

bearing or bending; also above certain L/d ratios, the load remains constant. Again, these tests showed that by using mild steel plates where the load is parallel to the grain, a 25% increase in bolt capacity could be taken but this does not apply when the load is perpendicular to the grain. This increase is reflected in Clause 3.22.8.

Example 12.5

Using a 12.7 mm diameter bolt, find the medium term load capacity for the joint given in Fig. 12.6 when the load is applied parallel to the grain. Assume Group J3 timber.

Referring to Table 12.12 with thickness equal to 35 mm and load parallel to grain, the permissible basic load is given as 2.40 kN for a 12.7 mm bolt.

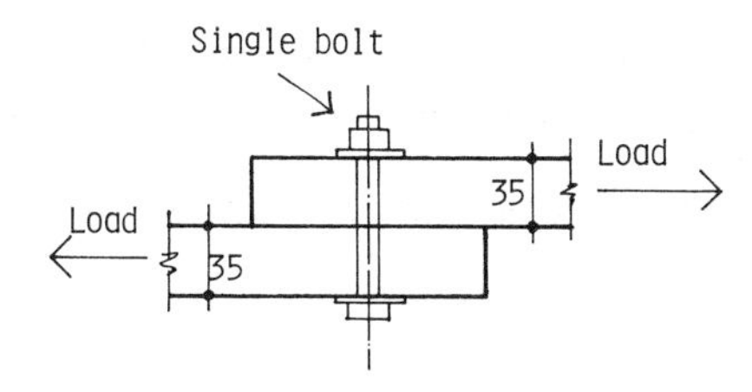

Fig. 12.6

As the load is of medium term duration then the increase factor of K_{21} applies and for bolts $K_{21} = K_{12}$.

for medium term $K_{21} = 1.25$

hence final joint capacity $= 2.40 \times 1.25 = 3.0$ kN

If the load had been applied perpendicular to the grain then it can be seen that the joint capacity would have reduced to

$1.34 \times 1.25 = 1.675$ kN

In the first part of this example (parallel), if one side of the joint had been replaced by a metal plate then the load capacity could have been increased by the factor of 1.25 but this would not apply to the second part (perpendicular).

In both cases, if the joint had been exposed the capacity would need to have been reduced by multiplying by 0.7, Clause 3.22.5. This same clause also states that where a member is jointed to a steel plate or jointed to another member by multiple bolts and is subject to drying after assembly, all basic loads must be reduced by multiplying by 0.4.

12.5.2 Loads at an angle to the grain

When a load can be said to be acting at an angle to the grain (common in truss design) then the load is given by the Hankinson formula

$$N = \frac{PQ}{P\sin^2\theta + Q\cos^2\theta}$$

where

θ = angle between direction of load and direction of grain
P = load parallel to grain given in Table 12.12
Q = load perpendicular to grain given in Table 12.12

Example 12.6

For the joint defined in Fig. 12.7, determine the limiting capacity assuming a 12 mm bolt in Group J2 timber and given medium term loading applies.

Limiting capacity for the tie member: From Table 12.12 load parallel, basic for 32 mm $=$ 2.50 kN

increase factor $K_{21} = 1.25$

joint capacity $= 2.50 \times 1.25 = 3.125$ kN

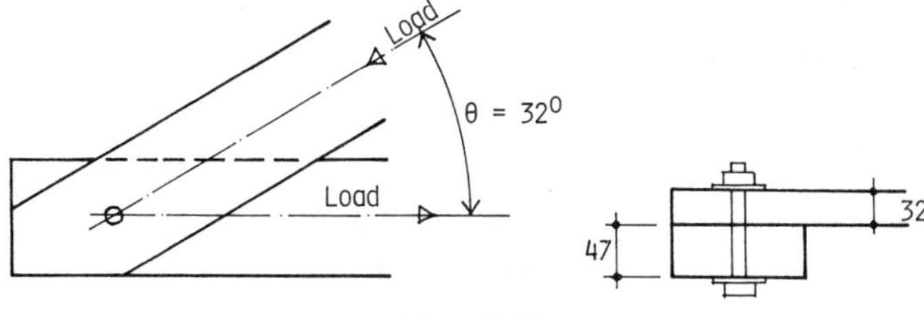

Fig. 12.7

Limiting capacity for inclined member: From Table 12.12 load parallel (P) for 47 mm = 2.50 kN, and load perpendicular (Q) for 47 mm = 2.28 kN. Therefore

$$N = \frac{2.5 \times 2.28}{2.5 \times \sin^2 32 + 2.28 \times \cos^2 32} = 2.43 \text{ kN}$$

increase factor $K_2 = 1.25$

joint capacity $= 2.43 \times 1.25 = 3.04$ kN

This example shows that even though the supported member is thicker, because of the inclination to the grain, its capacity is less than that of the tie and so it controls the design of the joint.

12.5.3 Multiple shear

In Clause 3.22.3 the code stipulates that for joints composed of more than two members, the basic bolt loads may be taken as the sum of the shear planes. A proviso to this is that any member having a shear plane on each side must have twice the tabulated thickness. Fig. 12.8 illustrates this proviso.

t_1 = tabulated thickness

$t_2 = t_1 \times 2$ (min.)

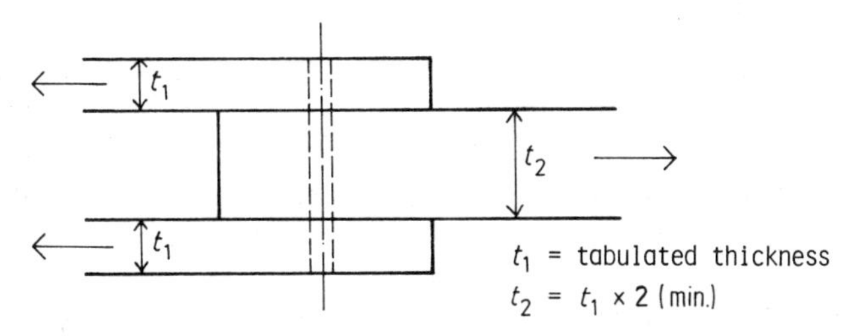

Fig. 12.8

All increase factors and limiting features apply as described for the single shear conditions.

Example 12.7

In the three member joints in Fig. 12.9, determine the limiting long term joint capacity using a 16 mm bolt in Group J3 timber

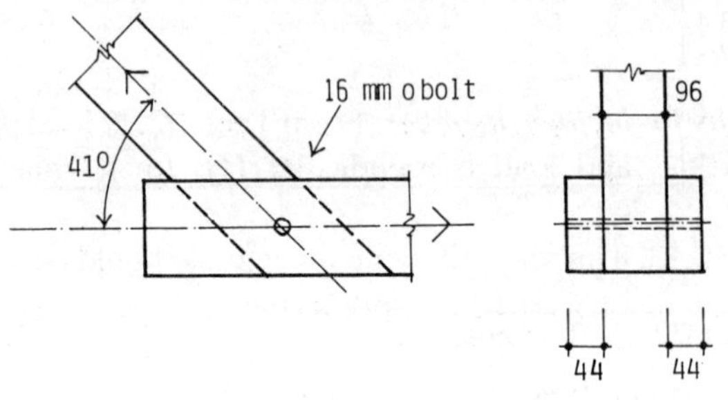

Fig. 12.9

Limiting capacity for the horizontal tie: From Table 12.12 load parallel to grain, for 44 mm = 3.77 kN

joint capacity = 3.77 × 2 shear faces = 7.54 kN

Limiting capacity for inclined tie: From Table 12.12 load parallel (P) for 96/2 mm = 3.79 kN and load perpendicular (Q) for 96/2 mm = 2.04 kN. Note that tabulated 47 mm is taken as the design thickness. Now

$$N = \frac{3.79 \times 2.04}{3.79 \times \sin^2 41 + 2.04 \times \cos^2 41} = 2.77 \text{ kN}$$

joint capacity = 2.77 × 2 = 5.54 kN

Therefore the joint capacity limit is controlled by the inner inclined tie.

12.5.4 Holes and washers

When considering the specification for bolted joints, the diameter of the hole should be as close to the diameter of the bolt as is practicable. Clause 5.4.3 of CP 112:Part 2, recommends that the hole should not be more than 1.6 mm larger than the bolt. Anything larger than this would lead to unacceptable joint slip. Bolts should not be positioned in any split which occurs close to or passes through the end of the timber section.

In this same clause, further recommendations are made as to the minimum size of washers to be placed beneath the head and nut of the bolt and these are

Table 12.13 Minimum sizes of washers

Diameter of bolt		*Minimum thickness of washer*		*Minimum side of square or diameter of washer*	
mm	*in*	*mm*	*in*	*mm*	*in*
9.5	3/8	3	1/8	51	2
12.7	1/2				
15.9	5/8				
19.0	3/4	5	3/16	64	5/2
22.2	7/8				
25.4	1				
28.6	9/8	6	1/4	76	3
31.8	5/4				

listed in Table 12.13. Timber to timber connections should never be bolt fixed without the benefit of washers to spread the bearing loads on tightening. Overtightening should be avoided at all times; similarly, consideration in the site application should be given to tightening bolts in trusses and structural frames after a short period of around eight weeks and again after, say, six months.

12.5.5 Spacing

To avoid splitting and to give sufficient timber for the transference of loads between bolts and toward ends and edges, the bolts should be properly

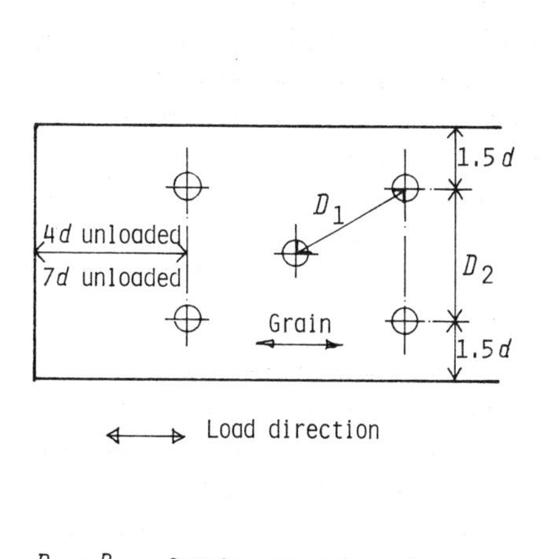

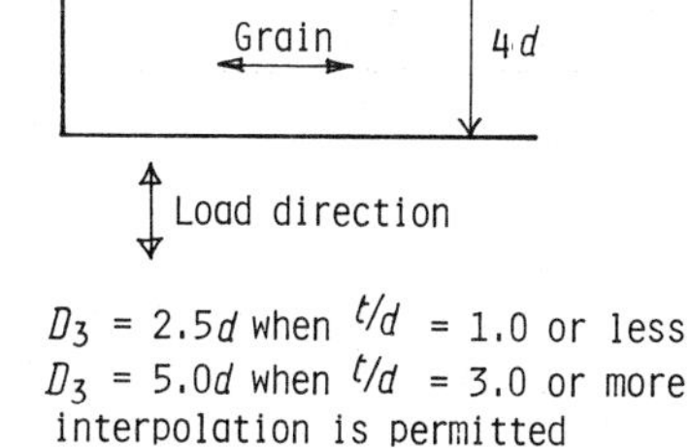

$D_1 = D_2 = 2d$ min. or washer size
d = diameter of bolt
t = thickness of timber

$D_3 = 2.5d$ when $t/d = 1.0$ or less
$D_3 = 5.0d$ when $t/d = 3.0$ or more
interpolation is permitted

Fig. 12.10

spaced. Clause 3.22.7 describes the limitations having due regard to the direction of the grain and the direction of the load. These are best represented by the diagrams in Fig. 12.10. If a load is applied at an angle to the grain, then the applicable spacing is controlled by the greater of either the parallel or perpendicular requirements.

12.6 Tooth-plate, split-ring and shear-plate connections

12.6.1 General descriptions

Of the three connections mentioned, split-rings and shear-plates are used for predominantly heavy duty loads or where demounting may be needed. They require special machined grooves and care and accuracy in assembly. Consequently, it is not surprising that the more commonly used connection is the tooth-plate where no special tooling is required other than high tensile steel drawing studs for embedding the teeth of the connectors into the timbers. Tooth-plates are not capable of carrying the same magnitude of load as split-rings and shear-plates.

The British Standard relevant to all three connections is BS 1579:1960 and the diagrams which follow indicate the general shape of each:

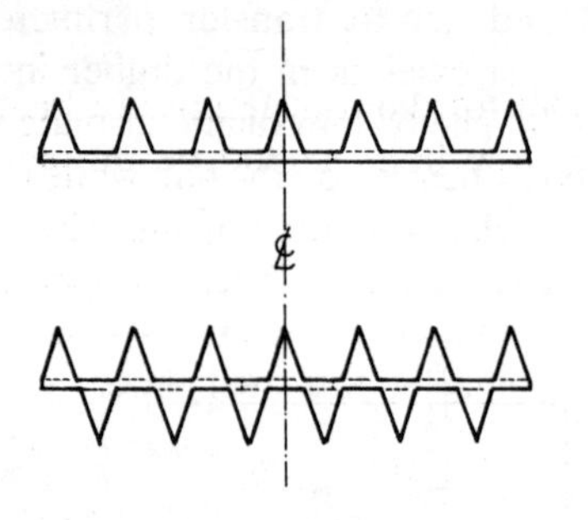

Single-sided round or square, each in sizes of 38, 51, 64 and 76 mm.

Double-sided round or square, each in sizes of 38, 51, 64 and 76 mm. They work in conjunction with bolts and washers and the hole allows tolerance, therefore, joint slip always occurs. The British Standard requires anti-corrosion treatment and more commonly this is sherardizing. The correct size of bolt must be used as detailed in CP 112:Part 2.

Split-rings may be bevelled or straight sided. A circular groove of half the depth of the ring (*D*) is cut out of each piece of timber and the ring is sprung into this groove thus transferring shear forces across the ring between members. They are provided in two sizes of 64 mm and 104 mm internal diameter and require, respectively, 13 mm and 19 mm diameter bolts, or the larger bolt if both rings are used on the same bolt. Because of the joint in the split-ring it is susceptible to moisture movement in the joint and so it is good where shrinkage or swelling may occur. They must be anti-corrosion treated.

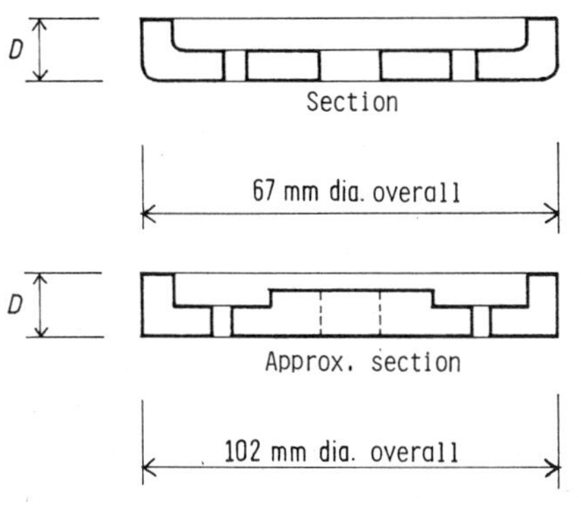

Pressed-steel shear plate (anti-corrosion treated).

Malleable cast iron shear plate (anti-corrosion treated). Shear plates require a specially cut round recess and are let into the timber to their full depths (D). In both cases they are used in conjunction with 19 mm diameter bolts and act to transfer perimeter bearing stresses from the timber into the bolt. The 67 mm diameter plate is limited by bearing on the bolt while the 102 mm plate is limited by shear in the bolt. They are used back to back in timber to timber connections or singly when connected to a steel plate.

12.6.2 Basic loads

Basic loads in softwood for each of the connection units previously described are given in CP 112:Part 2 with a modification by the Hankinson formula (see Section 12.5.2) to adjust for angle of load to grain. Basic loads are given for three species groups, namely J2, J3 and J4 for parallel and perpendicular to the grain.

It is considered beyond the scope of this book to enter into a detailed discussion on the various connector usage designs because of the protracted length of the subject. Therefore, none of the basic loads or modifications factors has been reproduced; however, it is suggested that the student examines the code of practice carefully against the following guide notes. In addition, a design example will be given indicating how a joint making use of toothed-plate connectors may be assessed. The principles may be followed for joints using split-rings and shear-plates.

12.6.3 Permissible connection loads

The permissible load is obtained by multiplying the basic load (using the Hankinson formula if necessary) by certain modification factors as follows:

$$\text{permissible load} = \text{basic load} \times K_{22} \times K_{23} \times K_{24}$$

In the case of shear-plates, their permissible value must never exceed values given in Table 32 of the code.

12.6.4 Modification factors

K_{22} is taken for the duration of the load either long term, medium term or short term and is equal to $(1 + K_{12})/2$ for toothed plates where $K_{12} = 1.0, 1.25$ or 1.50 for these term loads, respectively.

In the case of shear-plates $K_{22} = K_{12}$ provided that the resulting load does not exceed those given in Table 32 (see Section 12.6.3). K_{23} is a reduction factor of 0.7 for conditions related to long term exposure, for example in bridges, feet of exposed columns and the like. K_{24} is a modification factor which allows for sub-standard end and edge distances and spacing.

For the full design load to apply, these distance limitations must comply with those given in the various tables in the code; however, it is recognized in joint design that this is not always feasible and so sub-standards are allowed down to an absolute minimum beyond which it is not permitted to go.

These sub-standards relate to end distance (loaded and unloaded), edge distance (loaded and unloaded), and spacing (centre to centre of connectors). They are classified as

K_C = modification for end distance
K_D = modification for edge distance
K_S = modification for spacing

The lowest of these sub-standard modification factors is taken as being the value for K_{24}. If so required, the two remaining factors may be further lowered to obtain a value equal to K_{24}. For intermediate values of end and edge distances and spacing, interpolation of these modification factors is permitted.

12.6.5 Design example

The following design example will serve to show how the code of practice is used to design a joint, Fig. 12.11(a), using 64 mm round toothed plate connectors in Group J3 timber in dry conditions and subjected to medium term loading with rafter 47 mm × 145 mm and tie 47 mm × 120 mm.

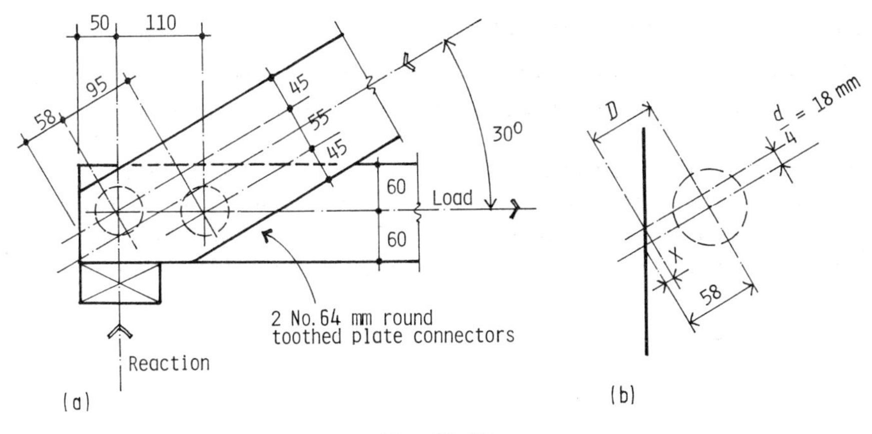

Fig. 12.11

By examining Fig. 12.11, it will be seen that the joint comprises two pieces of timber which cross each other touching face to face with one inclined to the other. Each member carries an axial load which must be capable of being transmitted in shear across the bolted connectors and into the other member. Clearly this generates two basic check conditions:

(a) inclined load into horizontal member
(b) horizontal load into inclined member.

Step 1

Determine the basic load for the connectors from Table 29 of CP 112:Part 2 and adjust for an angle to the grain of 30°.

Criteria = 63 mm round, connector one side, member thickness 47 mm
Group J3 timber

Hence by interpolating for the 47 mm thickness we have:

load parallel to grain (P) = 5.43 kN

load perpendicular to grain (Q) = 3.94 kN

Now adjusting by Hankinson $N = \dfrac{PQ}{P\sin^2\theta + Q\cos^2\theta}$

$$= \frac{5.43 \times 3.94}{5.43\sin^2 30 + 3.94\cos^2 30}$$

= 4.96 kN per connector

Step 2

For this basic load (N) to apply the connectors must be properly positioned in both members with regard to edge distance, end distance and spacing. Therefore we must now check for K_C, K_D and K_S in order to determine K_{24} and this may be tabulated as indicated in Table 12.14. However, before this table can be completed, it is necessary to check that the sub-standard end distance for the connector which is close to the splay cut (Fig. 12.11(b)) is not less than the limiting square cut end of 50 mm; to be in accordance with Clause 3.23.4.1. Now

$$x = 18 \tan 30° = \text{say } 10 \text{ mm}$$

giving

$$D = 58 - 10 = 48 \text{ mm} < 50 \text{ mm}$$

Therefore the splay cut end constitutes the limiting end distance. From this tabulation (Table 12.14), it can be seen that the loaded end distance modification factor governs giving $K_C = 0.68$ and therefore $K_{24} = 0.68$.

Table 12.14

	Referring to CP 112					
	Table 34		*Table 37*		*Table 40*	
	End distance		*Edge distance*		*Spacing*	
Member	Actual	K_C	Actual	K_D	Actual	K_S
Inclined into horizontal	48 (loaded)	0.68	65 min. = 37 mm	1.0	110 ($\theta = 30°, \phi = 30°$)	1.0
Horizontal into inclined	48 (unloaded)	0.95	45 min. = 37 mm	1.0	110 ($\theta = 30°, \phi = 30°$)	1.0

Step 3

Determine modification factors K_{22} and K_{23} and from these derive the permissible load and hence the joint resistance.

$$K_{22} \text{ for medium term loading} = \frac{1 + K_{12}}{2}$$

$$= \frac{1 + 1.25}{2} = 1.125$$

$$K_{23} \text{ take as dry} = 1.0$$

Hence

$$\text{permissible load} = \text{basic } (N) \times K_{22} \times K_{23} \times K_{24}$$
$$= 4.96 \times 1.125 \times 1.0 \times 0.68$$
$$= 3.80 \text{ kN per connector}$$

Therefore

$$\text{joint resistance} = 3.80 \times 2 \text{ No.} = 7.60 \text{ kN}$$

A point which requires a little further clarification is where the tables for K_S refer to the values of θ and ϕ.

θ defines the angle of the load to the grain direction and requires no further explanation but ϕ which is given as the angle of the connector axis to the grain direction is not as straightforward and, initially, this may be somewhat puzzling. However, it is best described as the line which joins the centre points of two or more connectors against which the grain direction should be measured. For example, by extracting the inclined member from Fig. 12.11(a), it can be seen that the connector axis ϕ is horizontal (0°) while the angle of the grain relative to this axis (θ) is inclined at 30° (Fig. 12.12).

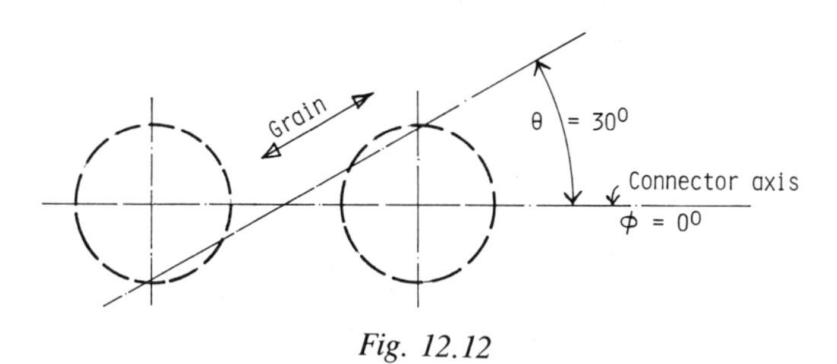

Fig. 12.12

12.7 Adhesives

12.7.1 Introduction

Adhesives for structural use are many and varied and the market choice for the user is wide with a good deal of research testing back up. In the main, glues used for structural purposes should satisfy the requirements of BS 1204:Part 1:1979 or BS 1444:1970.

Modern synthetic glues include those termed 'gap filling' which are extensively used by the manufacturer of structural components to fill small end gaps and the like in jointing features. The maximum permissible gap is recognized to be 1.3 mm as specified in BS 1204.

12.7.2 Performance classification

The performance of glues, particularly in relation to durability, must be known before a proper specification can be achieved by the designer. BS 1204 classifies glues by durability performance in the following four categories:

(a) Type WBP which are the initial abbreviations for weather and boil proof. This glue is highly resistant to varying degrees of weather exposure, sun, rain, cold, etc.
(b) Type BR short for boil resistant. Good weather resistance but fails after prolonged exposure.
(c) Type MR meaning moisture resistant. May be used to withstand moisture for a limited period only and, consequently, they are never specified in areas where prolonged exposure is intended.
(d) Type INT designated interior. Very rarely used by structural designers as it is only suitable for interior conditions of a good protected kind.

Of these four types described, by far the most extensively used in structural design is the Type WBP where RF and PF adhesives are used. BRE Digest No. 175 (1975) produced by the Princes Risborough Laboratory is entitled 'Choice of glues for wood' and gives guidance on exposure conditions and the type of the adhesive to be used. Table 12.15 gives this choice of adhesives and is reproduced from Digest No. 175 by permission of PRL.

12.7.3 Shear stresses

All glued joints are designed to resist shear, and other secondary stresses such as tension across the glue face should be avoided. In order to obtain this shear line, the glue spread should be as thin as possible while realizing the full potential of the adhesive. Clause 5.4.7 of CP 112:Part 2:1971 gives guidelines on the manufacturing process which should be used to obtain the full shear performance anticipated in the design.

The permissible shear stress in the glue line is that which is appropriate to the shear stress parallel to the grain given for the timber grade to be used. Clause 3.24.1 of CP 112 further describes the permissible shear for a load at an angle to the grain as being that which is obtained by applying the Hankinson formula of:

$$N = \frac{PQ}{P\sin^2\theta + Q\cos^2\theta}$$

where

θ = angle between direction of the load and the direction of the grain in one or more pieces of timber making the joint
P = appropriate grade stress for shear parallel to grain
Q = appropriate grade stress for shear at right-angles to the grain = $P/4$

Table 12.15 Choice of adhesives for different applications

Exposure category	*Exposure conditions*	*Examples*	*Recommended adhesive (see key)*	*Performance type*	*BS reference*
Exterior					
High hazard	Full exposure to weather	Marine and other exterior structures. Exterior components or assemblies where the glueline is exposed to the weather	RF	WBP	BS 1204:1979 Part 1
			RF/PF	WBP	
			PF (cold-setting)	WBP	
Low hazard	Exposed to weather but protected from sun and rain	Inside the roofs of open sheds and porches	RF; RF/PF; PF	WBP	BS 1204:1979 Part 1
			MF/UF[3]	BR	
Interior					
High hazard	In closed buildings with warm and damp conditions where a moisture content of 18% is exceeded and where the glueline temperature can exceed 50°C	Laundries; roof spaces	RF; RF/PF; PF	WBP	BS 1204:1979 Part 1
			MF/UF[3]	BR	
Low hazard	Heated and ventilated buildings where the moisture content of the wood will not exceed 18% and the glueline will remain below 50°C	Inside dwelling houses, heated buildings, halls and churches	RF; RF/PF; PF	WBP	BS 1204:1979 Part 1
			MF/UF[3]	BR	
			UF[4]	MR	
			Casein		BS 1444:1970 (type A)
Special	Chemically polluted atmospheres	Structures in the neighbourhood of chemical plants or associated with the manufacture of electrical batteries. Dyeworks. Swimming baths	RF; RF/PF; PF	WBP	BS 1204:1979 Part 1

MF/UF melamine/urea-formaldehyde
PF phenol-formaldehyde
RF resorcinol-formaldehyde
RF/PF resorcinol/phenol formaldehyde
UF urea-formaldehyde

It should be noted that this formula in general applies to jointing timber to timber and although Clause 3.24.2 of the code recommends the use of the same formula when using nail pressure for jointing plywood, it is suggested that the results should always be compared with the value given for rolling shear in the ply (Section 1.3.2). This is because there is no guarantee that the face glue line shear could be developed throughout the plywood thickness without causing rolling shear between the laminates. Stress concentrations commonly occur in the use of plywood at, for example, the junction between ply and chords in box beams and for these conditions Clause 4.6.3 of CP 112 requires the permissible shear stress to be reduced by 0.5. This should always be used unless there are sufficient acceptable test data permitting higher concentrated stresses to be adopted.

Example 12.8

In Fig. 12.13 if we give face contact widths of 72 mm in SS grade timber, the long term strength of the joint for Whitewood may be determined as follows:

$$N = \frac{PQ}{P\sin^2\theta + Q\cos^2\theta}$$

$$P = 0.86 \text{ N/mm}^2 \text{ (Table 3.7)}$$

$$Q = \frac{0.86}{4} = 0.215 \text{ N/mm}^2$$

$$\theta = 41°$$

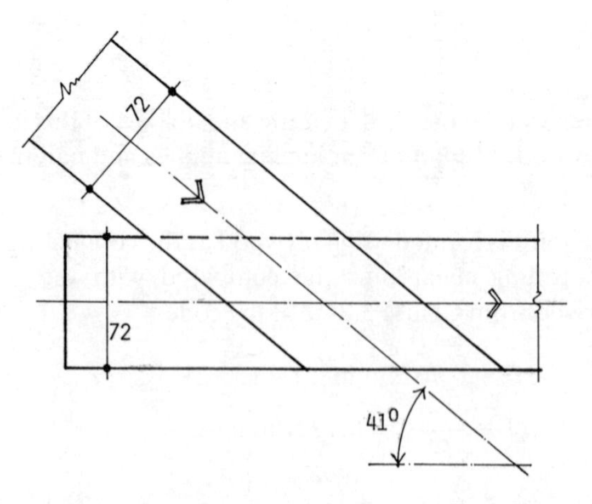

Fig. 12.13

giving

$$N = \frac{0.86 \times 0.215}{0.86 \times \sin^2 41 + 0.215 \cos^2 41}$$

$$= 0.375 \text{ N/mm}^2$$

$$\text{contact area} = \frac{72}{\sin 41°} \times 72 = 7901 \text{ mm}^2$$

$$\text{joint resistance} = 0.375 \times 7.901 = 2.96 \text{ kN}$$

Example 12.9

For the gusset joint given in Fig. 12.14 determine the contact areas required for the rafter and tie assuming numerical 50 grade Redwood is used in conjunction with Douglas Fir plywood. It may be assumed that the vertical

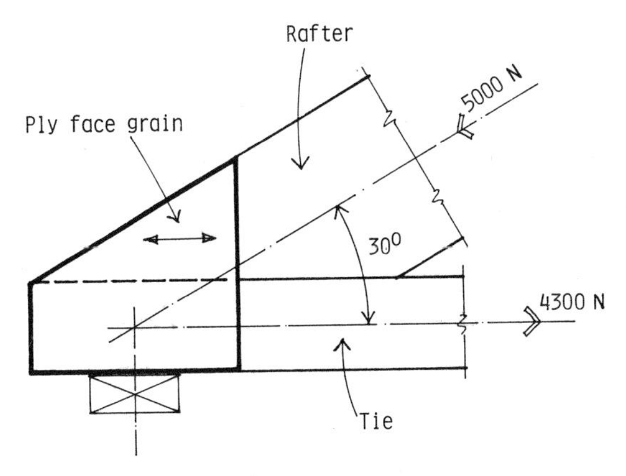

Fig. 12.14

reaction is resisted by the tie in bearing and no part of this load is transmitted into the plywood. Medium term loading applies and nail pressure loading is used.

Because the joint is formed using plywood as the connecting gusset plate, the permissible rolling shear must be compared with the permissible shear stresses derived from Clause 3.24.1 of the code

Rafter: $P = 0.76 \text{ N/mm}^2$ (Table 3.2)

$$Q = \frac{0.76}{4} = 0.19 \text{ N/mm}^2$$

$$N = \frac{0.76 \times 0.19}{0.76 \times \sin^2 30° + 0.19 \cos^2 30°} = 0.434 \text{ N/mm}^2$$

To comply with Clause 3.24.2 this stress must be reduced by multiplying by 0.9 because nail pressure is used, giving a final permissible shear stress of

$$N = 0.434 \times 0.90 = 0.390 \text{ N/mm}^2$$

By comparison with the rolling shear for Douglas Fir plywood, it is found that this permissible stress is 0.345 N/mm^2 which is smaller than the value for N and so must be taken as the limiting shear stress. Therefore contact area required

$$= \frac{5000}{0.345 \times 10^2} = 145 \text{ cm}^2$$

It should be noted that even though the applied forces are defined as of medium term duration, term loading increase factors are not used in the design of glued joints.

Tie: In the case of the tie it can be seen that the face grain of the ply is parallel to both the load and the grain of the timber with no angle of load to the grain. Consequently the full grade stress for the timber applies. However, once again, this is compared with the rolling shear and so it will be seen that:

$$N = 0.76 \times 0.9 = 0.684 \text{ N/mm}^2 > 0.345 \text{ N/mm}^2$$

giving

$$\text{limiting } N = 0.345 \text{ N/mm}^2$$

$$\text{contact area required} = \frac{4300}{0.345 \times 10^2} = 125 \text{ cm}^2$$

The code of practice, CP 112, Clause 3.17, recommends that in triangulated frameworks, due account should be taken of secondary stresses caused by such items as timber shrinkage, fixity and eccentricity. In the case of shrinkage, it is unlikely that much will occur because of the desirable low moisture content of around 15% when using adhesives. However, glued gusset joints will have a tendency to cause joint fixity and some degree of secondary stress concentration can occur, whereas eccentricity is a question of mathematical assessment.

Some designers adopt the policy of using the rolling shear and reducing this by 50% to allow for these unknowns, which under normal circumstances would tend to be a conservative approach. In the event, it is for the designer to judge, in regard to the size of the joints, the possible magnitude of fixity occurring and the occasion of eccentricity.

12.7.4 Pressure by nailing

In the previous examples it has been shown how the shear stress is reduced when the glue line pressure is obtained via nails.

CP 112, Clause 3.24.2 defines the length and spacing of the nails in the following manner:

(a) The length of the nails $\not< 4 \times t$ where t = ply thickness in contact with nail head.
(b) Spacing along the grain $\not>$100 mm for t = 10 mm or less and $\not>$150 mm for $t >$ 10 mm.
(c) Side spacing between lines of nails $\not>$100 mm.

In addition the general positioning of nails should be not less than those given in Table 24 of the code to avoid undue splitting of the wood, and staggering is recommended to obtain more even pressure.

Nailing is not considered to give more strength to the joint and consequently the nail strengths must be ignored in the design. Other means of applying the pressure may be used such as screws, improved nails and staples provided that they can be shown to give the equivalent holding pressure that is obtained by adopting the nailing procedures in Clause 3.24.2.

From the point of view of detail, very often the nail pressure spacing rules will determine the physical size of the gusset plate and it is not enough to define the contact area in the final design without examining the nail pressure requirement for each member forming the joint.

12.8 Integral toothed-plate connectors

12.8.1 Development

The original development of the integral toothed-plate connector took place in the USA and its subsequent introduction into this country, in the mid-1960s, made a remarkable and somewhat historical change in the building industry's treatment of roof construction. Before this time most roofing, using timber components, had been formed by the more traditional means of cutting and fitting each member *in situ*. With the development of this kind of connector the production and use of prefabricated roof trusses expanded rapidly until today the majority of roofs constructed from timber are formed by using the roof truss. Many different configurations are possible and their engineering solutions are in the main obtained by computer analysis.

12.8.2 Agrément certification

The concept of the plate is simple but its design is complex and such features as correct metal thickness, tooth shape, tooth orientation, tooth length and not least the ability to produce the tooled pressing die, cause many problems which must be solved before the connectors may be mass-produced. Final proof of the ability of the plate to perform satisfactorily in conjunction with the timber pieces to be joined, is provided by the British Board of Agrément where extensive testing is undertaken prior to the issue of their certification.

No design should be undertaken without the connector possessing design loads controlled by a current design certificate. These loads are given for use with various timber species and relate to the duration of the loading, the angle of the load to the grain direction and the angle of the load to the plate length (tooth direction).

12.8.3 Plate connection systems

There are several major suppliers of these connecting plates in the United Kingdom and they are known as 'system suppliers'. Choice of make is normally governed by cost, reliability of delivery, and the structural design support service offered. Competition is keen and it is not unusual for a customer to examine other systems and indeed at times to use more than one type of connector. Nearly all systems carry sophisticated computer software to support their product and the larger systems utilize the microcomputer for their customers' in-house use.

12.8.4 International Truss Plate Association

The International Truss Plate Association, commonly referred to as the ITPA, have done much to promote the proper use and specification of both the connectors and the roof truss. Notably they have from time to time produced technical literature in the form of bulletins highlighting important features of construction such as minimum standards for domestic roof bracing, water tank support platforms, erection sequence, etc. Included in their publications are recommendations for the correct procedure to be adopted in designing truss joints when using these specialist connectors. It is these procedures which we will use to show how two common joints of the FINK profile truss are designed, namely the heel joint and the third point joint in the ceiling tie.

12.8.5 Basic tooth loads

Before a design can commence we must have a suitable plate and be given the British Board of Agrément certified design tooth loads. By kind permission of Hydro-Air International Limited, a major system supplier (see Section 12.8.3), we shall use the Hydro-Nail 18 gauge E-plate covered by their current Agrément Certificate No. 80/720 from which their permissible design tooth

Table 12.16 Permissible load per nail (N) for Hydro-Nail 18 Gauge E timber fasteners in planed and sawn timber

Timber species	*Angle of load to fastener length direction,* $\alpha°$	*Angle of load to grain of member,* $\beta°$						
		0°	*15°*	*30°*	*45°*	*60°*	*75°*	*90°*
Long-term loading								
Commercial Western Hemlock	0	153	140	113	89	73	65	63
	15	145	133	109	88	73	65	63
	30	138	128	106	86	72	65	63
	45	130	121	102	85	72	65	63
	60	122	115	99	83	71	65	63
	75	111	105	93	80	70	64	63
	90	100	96	87	77	69	64	63
European Redwood, Whitewood,	0	126	117	98	80	68	61	59
Eastern Canadian Spruce	15	120	112	95	79	67	61	59
	30	114	107	92	77	67	60	59
	45	107	101	89	76	66	60	59
	60	101	96	85	74	65	60	59
	75	92	88	80	72	64	60	59
	90	82	80	75	68	63	60	59
Western White Spruce	0	99	94	83	71	62	57	56
	15	94	90	80	70	62	57	56
	30	89	86	77	68	61	57	56
	45	84	81	74	67	60	57	56
	60	79	77	72	65	60	56	56
	75	72	71	67	63	59	56	56
	90	64	64	62	60	57	56	56
Medium-term loading								
Commercial Western Hemlock	0	172	157	127	100	83	73	71
	15	164	150	123	99	82	73	71
	30	155	144	119	97	81	73	71
	45	146	136	115	95	81	73	71
	60	137	129	111	93	80	73	71
	75	125	119	105	90	79	72	71
	90	112	108	97	87	77	72	71
European Redwood, Whitewood,	0	142	132	110	90	76	68	66
Eastern Canadian Spruce	15	135	126	107	89	76	68	66
	30	128	121	104	87	75	68	66
	45	121	114	100	85	75	68	66
	60	113	108	96	84	73	68	66
	75	103	99	91	80	72	68	66
	90	93	90	84	77	71	67	66
Western White Spruce	0	112	106	93	80	70	64	63
	15	106	101	90	79	69	64	63
	30	101	97	87	77	69	64	63
	45	95	92	84	75	68	64	63
	60	89	87	80	73	68	64	63
	75	81	80	76	71	66	64	63
	90	72	72	69	67	64	63	63

Table 12.16 (continued)

Timber species	*Angle of load to fastener length direction, α°*	*Angle of load to grain of member, β°*						
		0°	*15°*	*30°*	*45°*	*60°*	*75°*	*90°*
Short-term loading								
Commercial Western Hemlock	0	191	174	141	111	92	82	78
	15	182	167	137	109	91	81	78
	30	172	160	132	108	91	81	78
	45	162	151	128	105	90	81	78
	60	153	143	123	103	89	81	78
	75	138	132	116	100	88	80	78
	90	125	120	109	96	86	80	78
European Redwood, Whitewood, Eastern Canadian Spruce	0	158	147	122	100	85	76	73
	15	150	140	119	98	84	76	73
	30	142	134	115	97	83	76	73
	45	134	127	111	94	83	76	73
	60	126	120	107	93	82	76	73
	75	114	110	101	89	80	75	73
	90	103	100	93	85	79	74	73
Western White Spruce	0	124	118	103	89	78	72	69
	15	118	112	100	87	77	71	69
	30	112	107	97	85	76	71	69
	45	105	102	93	83	76	71	69
	60	99	96	89	82	75	71	69
	75	90	89	84	78	73	71	69
	90	80	80	77	74	72	70	69

loads for long-term, medium-term and short-term duration are reproduced in Table 12.16.

Example 12.10

Referring to the design of the FINK truss in Example 10.6 we may obtain the axial loads for each of the truss members and proceed as follows.

Heel joint

Figure 12.15 gives the picture of the joint and indicates the position of the plate connector. The first step is to determine the values of $\alpha°$ and $\beta°$ in order that the permissible design tooth load may be obtained for each member, where:

α = angle of the load to the fastener length direction (sometimes referred to as the tooth direction)

β = angle of the load to the timber grain direction.

For the rafter force plate connection, $\alpha = 30°$ and $\beta = 0°$ giving, by reference to Table 12.16, a medium-term tooth load of 128 N/tooth using European

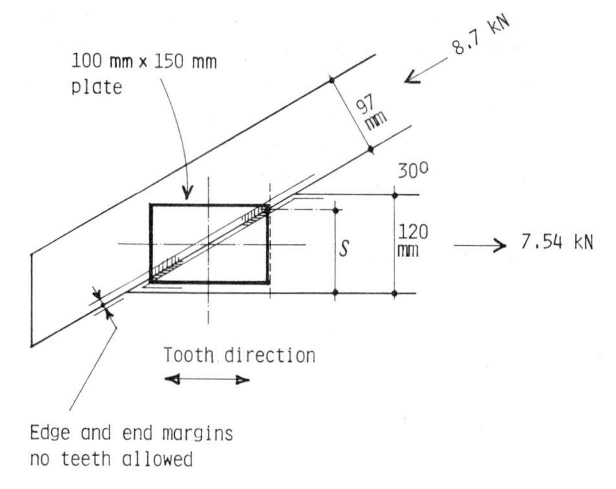

Fig. 12.15

Whitewood. Medium-term loading is chosen because the application of the theoretical snow loading creates the worst design condition.

$$\text{The applied axial load per plate} = \frac{8.70}{2} = 4.35 \text{ kN}$$

therefore

$$\text{the minimum number of teeth required} = \frac{4350}{128} = 34$$

Checking the ceiling tie force transference into the rafter we have

$$\alpha = 0°$$

and

$$\beta = 30°$$

giving a tooth load of 110 N/tooth.

$$\text{The applied load per plate} = \frac{7.54}{2} = 3.77 \text{ kN}$$

giving a minimum tooth requirement of

$$\frac{3770}{110} = 35$$

Therefore in this case it is the ceiling tie force, and its transference through the plate into the rafter, which controls the connecting area to the rafter.

The ceiling tie must similarly be checked for its own axial load connection and also for the transference of the rafter force across the joint into the tie. Clearly the tie's own connection does not control the design because its axial load is smaller than the rafter and because α and β are both 0°. Consequently when we examine the rafter force we find that for connecting to the tie we have $\alpha = 30°$ and $\beta = 30°$ giving a tooth load of 104 N.

The minimum tooth requirement is $\dfrac{3770}{104} = 37$.

To summarize, it is found that the minimum number of teeth required to each face of each member is:

rafter = 35 and tie = 37.

By the simple expedient of counting the teeth the correct plate size may be selected. However, in practice the design is normally based upon unit area and the computer makes the most economic selection from the given stock sizes.

Before either tooth count or unit area can be applied a reduction in the exposed surface area must be made to allow for the minimum end and edge distances given in the Agrément Certificate. In the former it is measured parallel to the grain direction from the line of cut, whilst in the latter it is taken perpendicular to the grain direction from the affected edge (refer to Fig. 12.16). In addition to these edge distances the ITPA stipulates that the lower edge of the plate must be a minimum of 3 mm above the soffit of the ceiling tie. This is to avoid both damage to ceiling finishes and to personnel whilst handling the truss.

The next step is to check the shear on the plate along the inclined interface between the rafter and tie caused by the action of the rafter force. The given plate size for this joint is 100 mm wide × 150 mm long, and by placing this

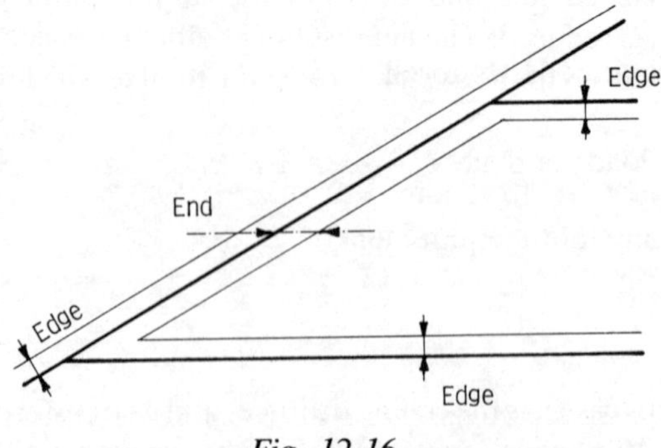

Fig. 12.16

symmetrically about the centre point of the splay cut we obtain the plating picture shown in Fig. 12.15. The total splay length through the plate depth is

$$\frac{2 \times 100}{\sin 30°} = 400 \text{ mm}.$$

For an angle of 30° to the plate length direction the certification allows a permissible shear force of 38 N/mm for all terms of load (Table 12.17). Hence the total inclined length required to satisfy the design would be:

$$\frac{\text{axial rafter load}}{\text{permissible shear force}} = \frac{8700}{38} = 229 \text{ mm}$$

as this is less than the length of 400 mm provided the plate meets the shear resisting standard.

Table 12.17 Shear strength. The maximum shear force acting on a fastener, for all three categories of load, must not exceed the values given below for the angle α, the angle between the fastener length direction and the direction in which the load is acting

Angle α	*N/mm of shear line*
0	31
15	34
20	38
35	38
45	46
60	55
75	55
90	46

Note: Values for intermediate angles can be interpolated.

Finally, from Fig. 12.15 it can be seen that when a broken line is drawn vertically along the extreme right-hand side of the plate it cuts the ceiling tie through the splayed line and consequently at this point the ceiling tie's effective depth is reduced. The net resulting timber cross-sectional area must therefore be checked for its ability to resist the total tensile force in the tie. By geometry:

$$S = 179 \times \tan 30° = 103.3 \text{ mm}$$
$$\text{Net area} = 35 \times 103.3 = 3615 \text{ mm}^2$$

For M50 timber

$$t_{\text{ppar}} \text{ (medium term)} = 4.6 \times 1.25 = 5.75 \text{ N/mm}^2$$

This may be increased by 1.1 for load-sharing conditions giving

$$5.75 \times 1.1 = 6.325 \text{ N/mm}^2.$$

Therefore

$$\text{tensile resistance} = \frac{3615}{1000} \times 6.325 = 22.86 \text{ kN}$$

which is clearly satisfactory.

Third point joint

Fig. 12.17 shows a typical method of approach to plating this particular joint with, this time, the plate oriented in the opposite direction to the heel joint. Once again permissible tooth values must be obtained for each member against their particular α and β slopes and Table 12.18 shows how the overall results may be tabulated.

For this particular joint the ITPA stipulates that the web members should be designed for 100% of the axial force against their appropriate tooth and grain orientation values, whilst the ceiling tie should be checked for the resultant net tensile force caused by T_1, T_2 and P (Fig. 12.18) where P is the panel load including tanks and man load if occurring. For this example P will be taken to include for a man load of 0.9 kN giving:

$$P = \text{man load} + \text{ceiling (dead + super)}$$

$$= 0.9 + 0.50 \times 0.6 \times \frac{8.4}{3} = 1.74 \text{ kN.}$$

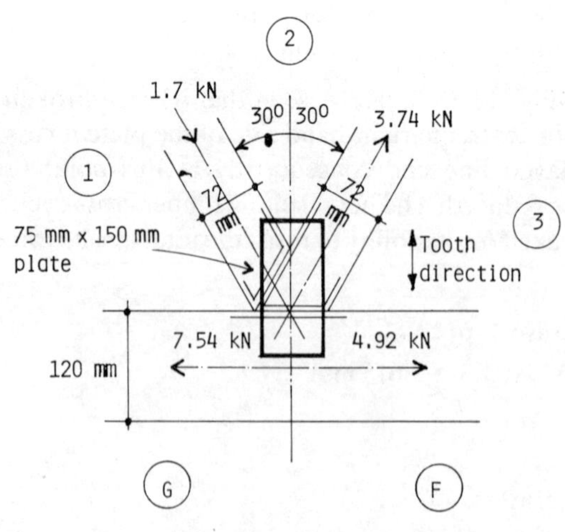

Fig. 12.17

Table 12.18

Member	*Axial load (W) (kN)*	$\alpha°$	$\beta°$	*Tooth value (t) (kN)*	*No. of teeth required per plate, W/2t*
1.2	1.70	30	0	0.128	7
2.3	3.74	30	0	0.128	15
GF	3.14	56.4	33.6	0.094	17

Resolving, we have

$$R = \frac{P}{\sin\theta_1}$$

$$\tan\theta_1 = \frac{P}{T_1 - T_2} = \frac{1.74}{7.54 - 4.92} = 0.664$$

$$\theta_1 = 33.60°$$

giving $\theta_2 = 56.4°$

Therefore

$$R = \frac{1.74}{\sin 33.60°} = 3.14\ \text{kN}.$$

Once again, checking for the plate's shear limitation, it can be seen that the residual force derived by subtracting T_1 and T_2 acts across the width of the plate to give the normal limit for design.

$$T_1 - T_2 = 7.54 - 4.92 = 2.62\ \text{kN}$$

$$\text{total plate width} = 2 \times 75 = 150\ \text{mm}$$

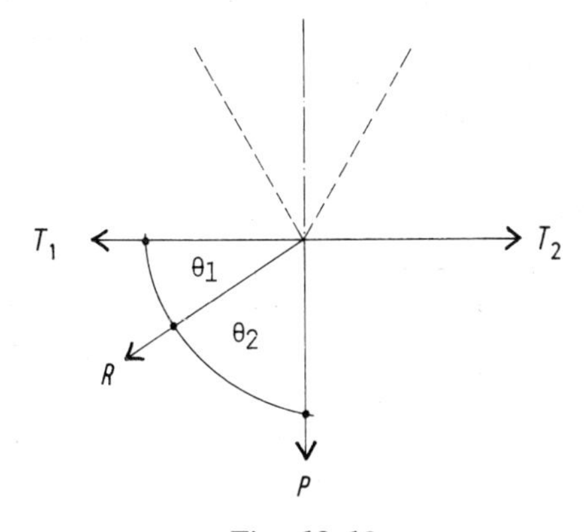

Fig. 12.18

From Table 12.17 when $\alpha = 90°$ the permissible shear force $= 46$ N/mm.

$$\text{Width required (per face)} = \frac{2620}{2 \times 46} = 28.5 \text{ mm}$$

The width provided is clearly satisfactory.

12.8.6 Eccentricity

Clause 3.17.1 of CP 112, Part 2:1971 concludes that where it is not practicable for the centroid lines of members to intersect at the joint, due account should be taken of the eccentricity. It will be noticed that in the preceding calculations eccentricity does occur but no account is taken of it. This is because all such plate connectors are determined by prototype testing of similar joints and eccentricity is automatically accounted for in the resulting certified values.

12.8.7 Plating tolerance

It is recognized that manufacturing techniques can lead to misplacement of the connector plate, in which case a joint's strength may be adversely affected. It is with this in mind that the ITPA makes the recommendation that all joints are considered, at the design stage, for possible displacement. In order to achieve this, a minimum value of 5 mm is set and each member within the joint must be considered for the worst possible misalignment in north, south, east and west directions.

12.9 Other jointing devices

In this chapter, an effort has been made to cover the salient features in those jointing devices which form the majority of usage. Many others do exist which it is beyond the scope of this book to enter into in detail; however, the following diagrams and notes will serve as a general guide to which the student may refer for further study:

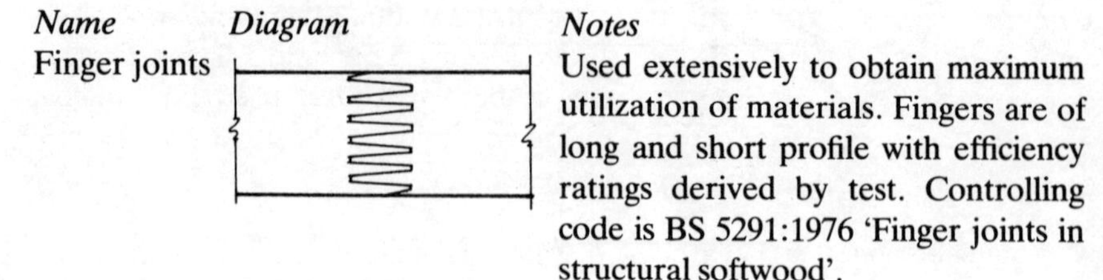

Name — *Diagram* — *Notes*

Finger joints — Used extensively to obtain maximum utilization of materials. Fingers are of long and short profile with efficiency ratings derived by test. Controlling code is BS 5291:1976 'Finger joints in structural softwood'.

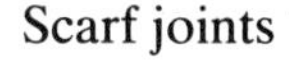

Scarf joints

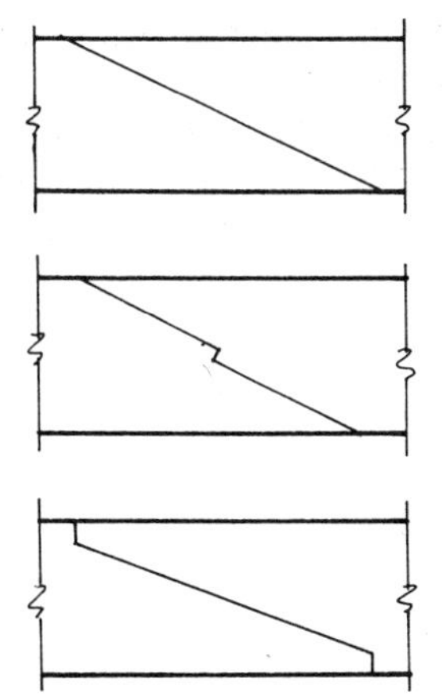

Slopes vary from 1 in 6 to 1 in 12 and CP 112 gives efficiency ratings for plain scarf joints used in glue-laminated construction. If joints are clamped by nailing the nail head must be recessed to avoid damage if surface preparation follows.

Framing anchors

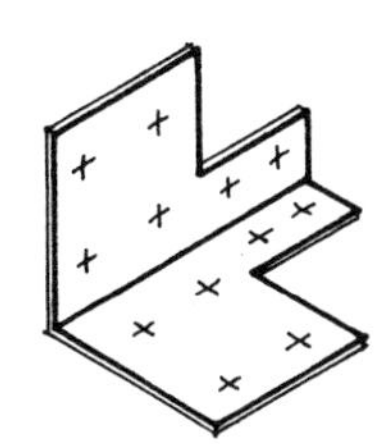

Cut and punched from thin pieces of galvanized metal. Used for joining pieces of timber at right-angles. Ordered by leg length and handed direction required.

Joist hangers

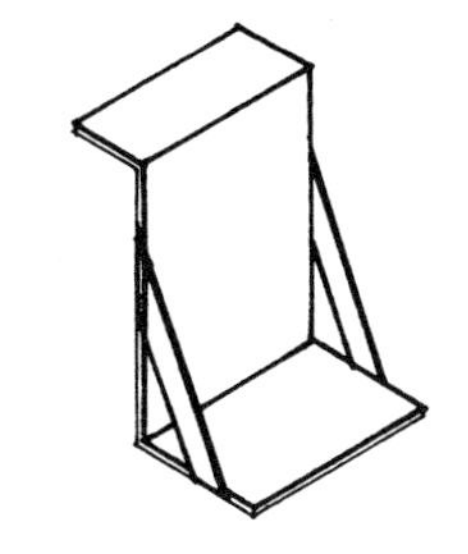

Shapes and choice are many and varied. Load capacities are given by the manufacturers having been obtained by prototype testing. Finishes are variously red oxide paint, galvanized or sherardized. Designers should be aware of such features as bearing, rotation and withdrawal from seatings.

Flat straps

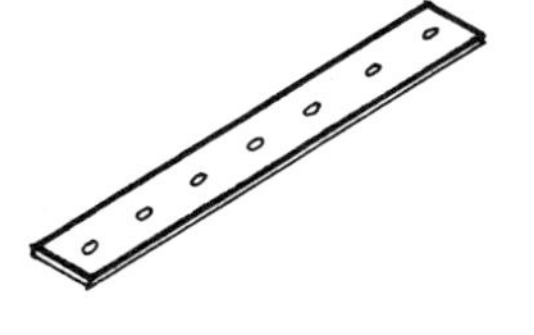

Proprietary galvanized with pre-punched holes at around 25 mm centres. Width 30 mm and two thicknesses of 2.5 mm and 5.0 mm. Can be pre-bent to order, used for holding down, jointing lengths and lateral restraint.

Pre-punched plates

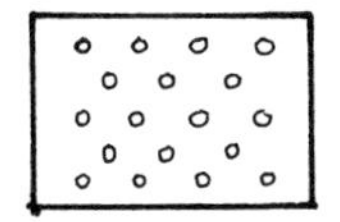

Flat galvanized plates with pre-punched holes used for end to end splicing, jointing trusses, etc.

Truss clips

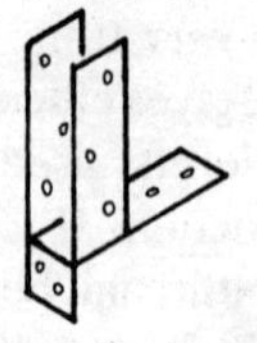

As the name implies used for holding down the ends of trusses over wall plates. Preferable to scew nail fixing which can cause damage. Thin galvanized metal strips bent to shape.

13
Good timber detailing

13.1 Observations

Good timber detailing can mean many things to different people. The lay public probably never appreciate good detailing simply because questions do not arise until problems occur but, conversely, bad detailing may cause user irritation. For example, hazardous protruding features, narrow impractical access ways, bulk-heads over stairs which obstruct passage of furniture, ill-fitting doors and windows caused by the movement of poorly seasoned timber, rotting end grain joints; these are but some of the features which would be evident to the layman.

On the other hand, an architect may well allow other features a greater degree of importance like obtaining good clean lines and pleasing grain, or neat unobtrusive jointing devices, or sound weathering details and so on.

For the timber engineer, most of the previous examples mentioned would not feature as highly in his thinking as would, say, the requirements for detailing a structural joint or the economics of timber species, or the consideration of chosen size for loads to be carried.

However, no matter the discipline involved, the important word is 'detail' for, in this word, lies the key to success or failure. The word implies special examination with an item by item approach to the subject matter and consideration of all salient points. Good detailing flows naturally from succinct and disciplined thought coupled with broad experience; the former may be quickly cultivated but the latter may only benefit from time. The wise student will quickly see that theoretical studies must always lead to practical application, and with the mastering of both, comes success. If, in any design discipline, one of these attributes is divorced from the other or the student allows one to predominate, then inevitably success will be that much harder to obtain. Consequently, good detailing, which is born of both, cannot be overstated.

13.2 Engineering detail

Inevitably, because this book concentrates on the design engineering of timber, our interpretation of good timber detail must relate more to the engineering aspect than to other auxiliaries. Notwithstanding this, one must not ignore the inter-relationship of other disciplines which may have some influence on the end result.

Bad detail will inevitably stem from hurried thought and a vagrant approach to a project; therefore, to obtain some degree of self-discipline, it is worth while pursuing a process which we would call 'Pre-detail think' and list some self-imposed questions.

13.2.1 Pre-detail think

(a) Have I understood fully the task in hand? If not, think again or seek more experienced advice.
(b) Is my approach the best solution or are there other ways? If so, I may need to examine more than one path and make comparisons.
(c) Does the detail I am about to employ satisfy any architectural restraints?
(d) Am I clear on the general specification and do I comply?
(e) Are the materials employed readily available? If not, what are the alternatives?
(f) Does the detail require a factory process and, if so, can it be made? If in doubt, consult the fabricator.
(g) Does the detail require an *in situ* application? If so, is it practical? Are there any site obstructions to consider? Does it require special tools? If in doubt, consult the builder.
(h) Have I given all relevant information to all concerned?
(j) Finally, but equally as important as all the previous questions, have I given my client what was requested, employing safety and economy efficiently and to the best of my ability or could I have done better?

It may well be that other, equally important questions occur to the reader, in which case the list will be extended. But, for the time being and to lend some substance to the subject, we will terminate the list here and re-examine some of the issues and enlarge upon them with examples.

Point (b)

A heavily loaded beam is indicated on the architect's drawing to support a pitched roof construction (Fig. 13.1). The detail is related to traditional construction and the designer's task is to convert the building to a structural timber frame supporting all loads.

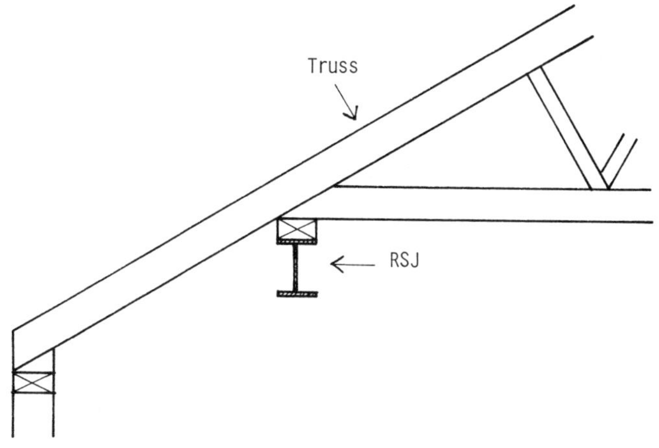

Fig. 13.1

One of the designer's details is to replace the steel beam with a flitch beam because the headroom restricts any increase in depth thereby restricting the use of solid sections of softwood or hardwood. Equally a ply box beam will not work. Is there an alternative which should have been examined? The answer is yes and Fig. 13.2 indicates an alternative position for a supporting beam. This position contains a number of advantages. It gives a greater depth zone and so increases the choice (perhaps a lattice beam?); it eliminates a down-stand beam which may better suit, architecturally; it does not involve the use of a secondary factory process such as steel fabrication; it makes possible the use of timber sections which are more readily available; it makes the end bearings more amenable to the overall construction concept, i.e. timber

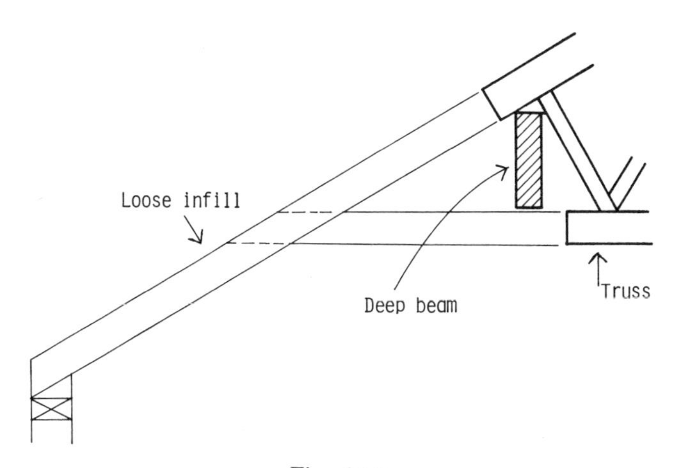

Fig. 13.2

orientated throughout. The designer should offer the two solutions for determination of all costs.

Point (e)

A designer has employed a ply box beam section to span 6.9 m overall supports. Structural calculations indicate the general cross-section with a clear specification of size and quality of materials. The architect's drawing indicates that the beam is to be supplied by a specialist manufacturer and, subsequently, the manufacturer receives the calculations. He is unable to comply with the design because he cannot obtain the solid cross-sections for the chords in the required overall lengths of 6.9 m.

This is a typical example of how lack of thought can lead to a poor specification and inevitable delays. The materials employed are not readily available, so what are the alternatives? The chords can be formed from continuous finger jointed thinner sections, glued together with the finger joints staggered. Or an alternative is to keep the solid section as specified and apply a suitable scarfed joint glued together to extend the usable length.

Point (f)

Poor detail and design in regard to this question is invariably a result of little if no appreciation of all of the particular factory processes.

The profile to the prefabricated roof truss indicated in Fig. 13.3 is issued for production by the designer and is rejected by the fabricator because it cannot be delivered. It is too tall to be stacked vertically on the lorry and too wide in span to be delivered flat without special support and escort. Had this general question been part of the designer's total check list, no doubt the restriction imposed by transport would have been spotted. The solution is a division of

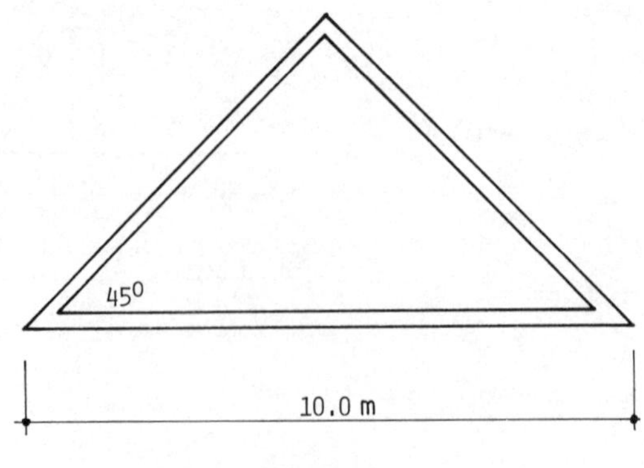

Fig. 13.3

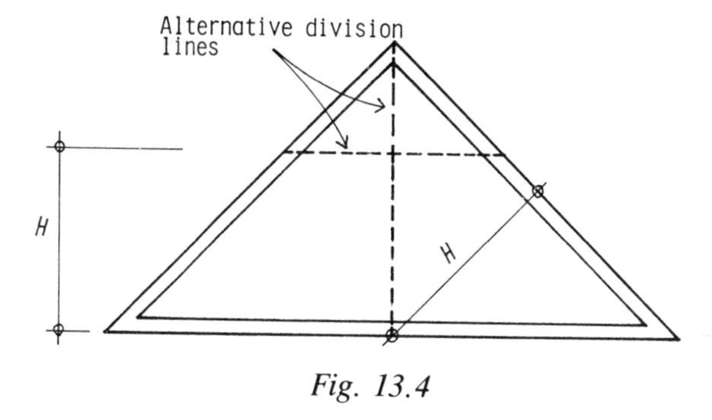

Fig. 13.4

the truss into two sections, either vertically or horizontally (Fig. 13.4), though the latter is preferable as it does not involve possibly complex site construction splice joints. Normally the limiting height restriction (*H*) will be taken to be around 4.0 m.

The plywood faced wall panel indicated in Fig. 13.5 is issued for manufacture and it is to be used in the upper floor gable wall to a domestic house. Checking against 'pre-think list' – can it be made? The answer is probably

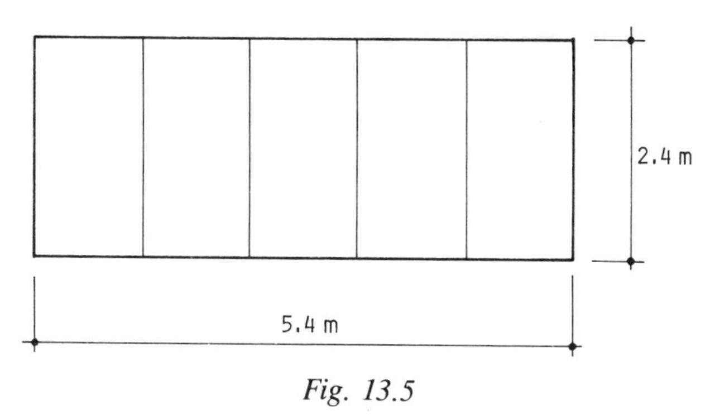

Fig. 13.5

'yes' but, both in the factory and on the site, crane handling will be required and this may be neither available nor desirable.

Point (g)

The roof of a narrow-fronted town house is to be converted into a habitable room space. The architect's drawing indicates support beams as shown in Fig. 13.6 with site location indicated in Fig. 13.7. The designer's beam size is 100 mm × 250 mm solid section. The designer's section is shown to be adequate but can the beams be installed? If the existing ceiling is to remain and the building process requires the installation of the beams prior to cutting for a new staircase, how are the beams to be installed? Here is a clear example of

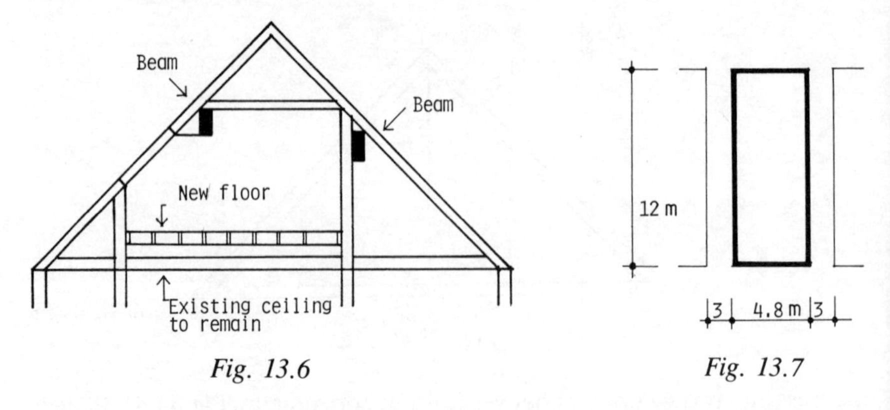

Fig. 13.6

Fig. 13.7

the necessity for the designer to consider the building process and to clarify his detail by previous consultation with the builder.

Point (h)

The purpose of detail is to inform and often poor detailing lacks proper information. The skilled detailer is able to give as much as is required and no more. No unnecessary frills or embellishments are used and there is no cluttering of information. It is a skill derived largely from years of practice and so it cannot be developed overnight or acquired by reading this book but, perhaps, the following guidelines will help:

- Apart from the written word, the detailer's principal asset should be the ability to communicate pictorially by means of sketches or scaled drawings.
- Practise the written word by imagining that you are the recipient and know nothing of the designer's thinking. Inform by using concise notes and simple everyday non-flowery words. If in doubt, write out the note first and précis as necessary until you are entirely satisfied.
- A detail will invariably form part of a whole feature of construction. This means that highlighting can be usefully employed to emphasize important features. This may be achieved by employing heavy lines or shading techniques.
- Always give salient dimensions so that the item can be made and fully located into its correct location. If fabrication is required, always prepare a separate drawing for this purpose alone.
- When giving timber cross-section sizes, these should always be the actual size required and a general note to this effect should be added to the drawing.
- Always give the timber species, grade required and moisture content. Similarly, clearly define quality of plywoods.

- Specify timber treatment if required.
- All bolts, nails, screws and special timber to timber connections should be clearly specified including treatment if required, e.g. sherardizing, electrogalvanizing, etc.
- Any tolerances of fit must be given.
- Do not forget to indicate pre-camber as and when required. This can often be omitted from fabrication drawings with disastrous results.
- If adhesives are to be used, it is important that they are covered by a specification note which should define the quality of the glue. If nail pressure is to be used, define the centre to centre spacing and the number of rows required plus the distance between rows and the minimum edge and end distances. Indicate that a squeezed glue line should be seen before applying any finishing techniques.
- Structural joints in timber are extremely important and details should be produced enlarged sufficiently to give good definition and to allow proper construction.
- Always cross-relate to relevant architectural drawings and specialist details.
- If a sketch or drawing is only intended to supply preliminary information, then say so by adding a bold note such as: PRELIMINARY INFORMATION ONLY – NOT TO BE USED FOR CONSTRUCTION.

These are a few guiding notes, by no means exhaustive, but sufficient to form a purposeful path to sound and workmanlike detail.

Finally, this observation is offered. Skilful detailing is largely a dying art, born of slow but inevitable changes within the building industry; changes in education, changes in training and, significantly, the remarkable explosion in the field of microcomputer technology. Such changes must be accepted if we are to move forward but it is suggested that we should never move forward if it is at the expense of losing skills which clearly should be retained and integrated with the new.

Therefore, new generations of budding engineers would be well advised to practise the art of the old where the drawing-board played an important role and the detailer's skill determined many a problem. For when the computer breaks down, we may be sure that our new breed will never be caught lacking as indeed the old guard rarely are.

14
Case study exercises

14.1 Introduction

The foregoing chapters have attempted to introduce the reader to the basics of timber engineering and to the materials commonly used. In addition some attempt has been made to explain how loading is assessed and how to express solutions in well-presented structural calculations. There have been many examples with a step-by-step approach to the answers and hopefully these have given the reader an insight into various ways of solving those problems which relate to elements of building which are common features of design.

In order to round off and to add some substance to the contents of this book it is hoped that the reader will undertake the tasks in this chapter, perhaps without first referring to the solutions which are offered. It is suggested that the answers should be presented in accordance with the recommendations of Chapter 8.

14.2 Tasks

The tasks which follow number eight in all and they relate to èlements of construction covered in this book. Some interpretation and thought will be necessary and it is not suggested that there are standard answers. But whatever the readers' solutions, it will be of interest to compare the results with those offered by the author.

SIMPLY SUPPORTED SOLID SECTION BEAM

Case Study 1 – Window lintel

ROOF SPECIFICATION:	Concrete tiles with standard loading as per CP 112,Part 3:1973. Roof trusses at 600 c/c with 35° slope.
TIMBER:	European Redwood or Whitewood
STRESS GRADE:	SS

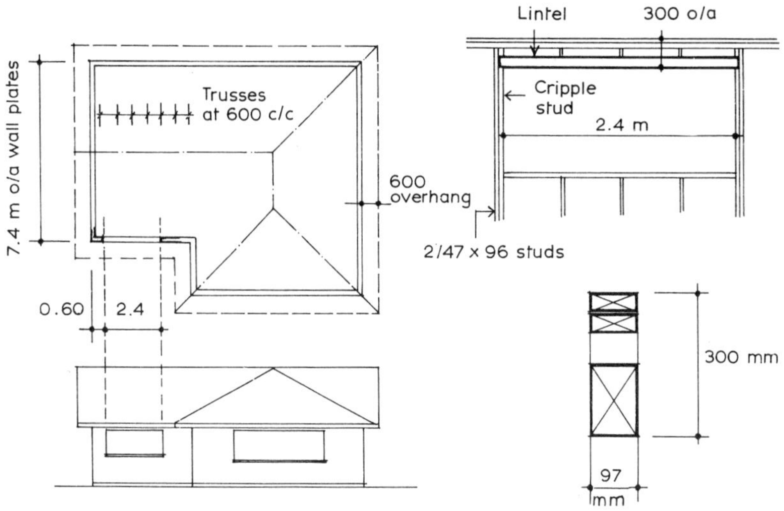

TASK:

1. Determine the design loading on the lintel.
2. Determine the minimum theoretical depth required and relate this to the nearest practicable timber depth.
3. Check the bending stress.
4. Check the bearing stress over the cripple stud.
5. Check the shear stress.
6. If the single solid size determined is not available, what are the alternatives?

Solution to Case Study 1

References

1. CP 112:Part 2:1971 'The structural use of timber'
2. CP 112:Part 3:1973 'Domestic roof trusses'
3. CP 3, Chapter V, Part 1:1967 'Dead and imposed loading'

General loading

Roof:

Super (0.75 − 0.09) for 35° slope $= 0.66$

Dead $\dfrac{0.685}{\cos 35°}$ (CP 112, Part 3) $= 0.84$

Ceiling: (Dead + Super) $= 0.50$

$2.00\ \text{kN/m}^2$

Lintel loading

P P P
= 600 600 =
2400 + 47
= 2447 c/c bearings

Load (P):

$$\text{Roof span} = 2.0 \times 0.60 \times \frac{7.4}{2} = 4.44$$

$$\text{Overhang} = 1.75 \times 0.60 \times 0.75 = 0.79$$

5.23 kN

From standard reference tables:

$$M = \frac{PL}{2} \qquad d = \frac{19PL^3}{384EI}$$

Depth required

The depth is determined by substituting the permissible deflection into the equation for the deflection formula and transposing to determine the moment of inertia as follows:

$$I_x = \frac{19 \times 5230 \times 2447^3}{384 \times 5700 \times 0.003 \times 2447} = 9061 \times 10^4\,\text{mm}^4$$

$$\text{from } I_x = \frac{bd^3}{12}$$

$$\text{then } d = \sqrt[3]{\left(\frac{9061 \times 10^4 \times 12}{97}\right)} = 224\,\text{mm}$$

Nearest practicable depth = 225
Checking against Clause 3.13.3 (Table 17) of CP 112, Part 2:1971

$$\text{then } \frac{d}{b} = \frac{225}{97} = 2.32 < 3 \text{ which is satisfactory.}$$

Check bending stress

$$M = \frac{5.23 \times 2.447}{2} = 6.40\,\text{kN m}$$

$$f_{\text{apar}} = \frac{M}{Z_x} \text{ where } Z_x \text{ for } 97 \times 225 = 818 \times 10^3\,\text{mm}^3$$

$$f_{\text{apar}} = \frac{6400}{818} = 7.82\,\text{N/mm}^2$$

$$f_{\text{ppar}} = 7.3 \times \underset{(K_{12})}{1.25} = 9.12\,\text{N/mm}^2 > 7.82 \qquad \text{OK}$$

Bearing stress over cripple stud

$$\text{Load over cripple stud} = P \times 2.5 \text{ (No. of trusses)}$$

$$= 5.23 \times 2.5 = 13.08 \text{ kN}$$

$$C_{\text{aperp}} = \frac{13\ 080}{47 \times 97} = 2.87 \text{ N/mm}^2$$

$$C_{\text{pperp}} = 1.55 \times 1.25 = 1.94 \text{ N/mm}^2 < 2.87 \qquad \text{NO!}$$

Therefore a second cripple stud would need to be installed.

Shear stress

$$\text{Shear load } V = P \times 1.5 \times \frac{3}{2} = 5.23 \times 1.5 \times \frac{3}{2} = 11.76 \text{ kN}$$

$$\text{Shear capacity } \bar{V} = \frac{0.86 \times 1.25 \times 97 \times 225}{1000}$$

$$= 23.46 \text{ kN} > 11.76 \qquad \text{OK}$$

Final practicable size

PROVIDE: 97 × 225 actual size solid timber section in European Whitewood to SS grade

Possible alternatives

1. Use of hardwood
2. Use of 2/47 × 225
3. Increase in grade stress and use of 2/38 thicknesses spaced apart.

Case Study 2 – Beam trimming floor opening

FLOOR SPECIFICATION: General office with boarded and joisted floor and ceiling of 12 mm plasterboard + skim. Directly above trimmer beams is a 50 × 100 at 600 c/c timber-framed stud wall 2.4 m high with 12 mm dry-lined plasterboard to inside face 100 mm glass fibre quilt insulation and 9 mm plywood external face.

TIMBER: European Whitewood

STRESS GRADE: SS

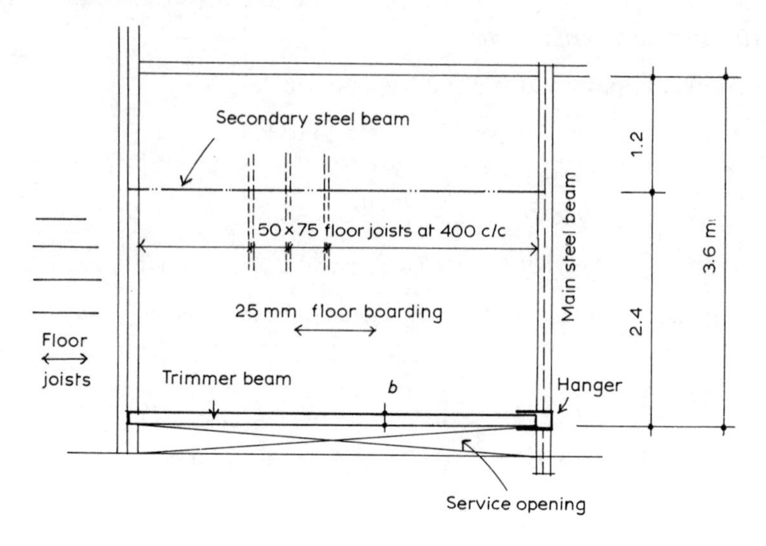

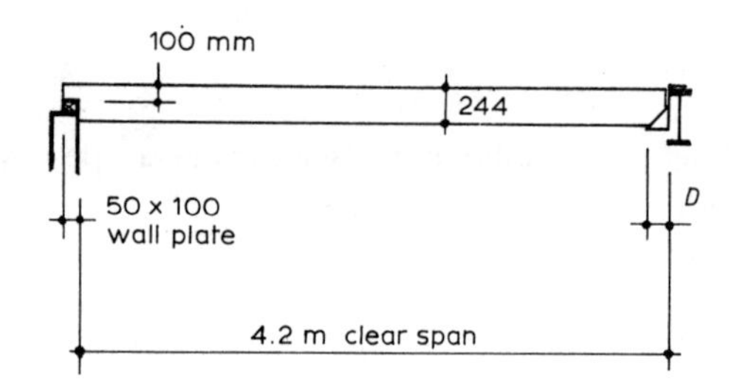

TASK:

1. Determine minimum theoretical width (b) required and relate this to most practicable solution.
2. Find minimum theoretical depth of seating (D) and relate to practicable hanger depth.
3. Check detail at left-hand support, giving any comments and suggestions.
4. Determine maximum bending stress.

Solution to Case Study 2

References

1. CP 112, Part 2:1971 'The structural use of timber'
2. CP 3, Chapter V, Part 1:1967 'Dead and imposed loading'

General loading

Floor:

Super (CP3 Chapter V, Part 1:1967)		2.50
25 mm floor boards	0.13	
50 × 175 floor joists at 400 c/c	0.11	
12 mm plasterboard + skim	0.18	0.42
		2.92
	say 3.0 kN/m^2	

Wall:

Plasterboard 12 mm	0.13
Framing 50 × 100 at 600 c/c	0.04
Plywood 9.0 mm	0.06
	0.23 kN/m^2

Beam loading

Take as simply supported with UDL

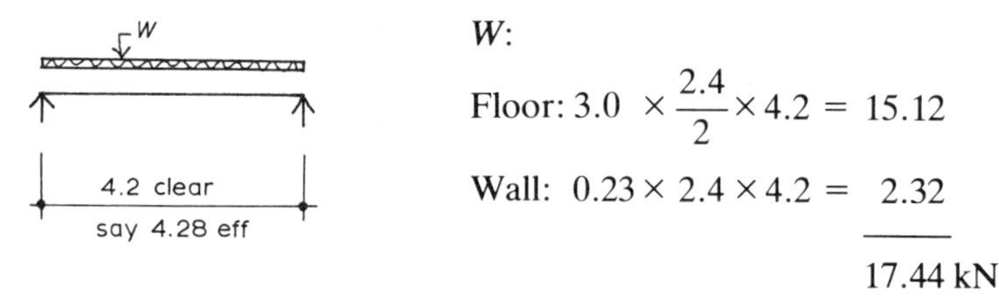

W:

Floor: $3.0 \times \dfrac{2.4}{2} \times 4.2 = 15.12$

Wall: $0.23 \times 2.4 \times 4.2 = 2.32$

17.44 kN

say 17.50 kN allowing for SW beam

From standard reference tables:

$$M = \frac{WL}{8} \text{ and } d = \frac{5WL^3}{384EI}$$

Width required

The width is determined by substituting the permissible deflection into the equation for the deflection formula and transposing to determine the moment of inertia as follows:

$$I_x = \frac{5 \times 17\,500 \times 4280^3}{384 \times 5700 \times 0.003 \times 4280} = 244 \times 10^6 \text{ mm}^4$$

$$\text{from } I_x = \frac{bd^3}{12}$$

$$\text{then } b = \frac{244 \times 10^6 \times 12}{244^3} = 202 \text{ mm}$$

The depth of 244 mm indicates that finished sizes are required and if 3 No. members bolted together are considered then the minimum practicable width is $72 \times 3 = 216$ mm.

Having decided on 3 No. members bolted together, it is now possible to reconsider the value of E by relating it to E_N where $N = 3$, $E_{mean} = 10\,000$ N/mm^2 and $E_{min.} = 5700$ N/mm^2

$$E_N = E_{mean} - \frac{E_{mean} - E_{min.}}{\sqrt{N}}$$

$$E_N = 10\,000 - \frac{10\,000 - 5700}{\sqrt{3}} = 7520 \text{ N/mm}^2$$

$$\text{therefore the modified } I_x = 244 \times \frac{5700}{7520} = 185 \times 10^6 \text{ mm}^4$$

$$\text{hence } b = \frac{185}{244} \times 202 = 154 \text{ mm}$$

This shows a considerable saving and by using 2/60 mm and 1/47 mm members we have a finished total width of 167 mm with 154 mm required.

Depth of seating (D) into hanger

$$\text{Beam reaction} = \frac{17.5}{2} = 8.75 \text{ kN}$$

$$C_{pperp} = 1.55 \text{ N/mm}^2$$

No increase factors are allowed because floor super loadings are regarded as long term; also the beam is a principal member and so load sharing does not apply.

$$\text{Bearing area required} = \frac{8750}{1.55} = 5646 \text{ mm}^2$$

The width is known to be 167 mm and so:

$$b = \frac{5646}{167} = 33.8 \text{ mm min.}$$

therefore to allow for an end cut tolerance, it is considered that 38 mm should be taken as an absolute minimum for the hanger depth.

Left-hand support detail

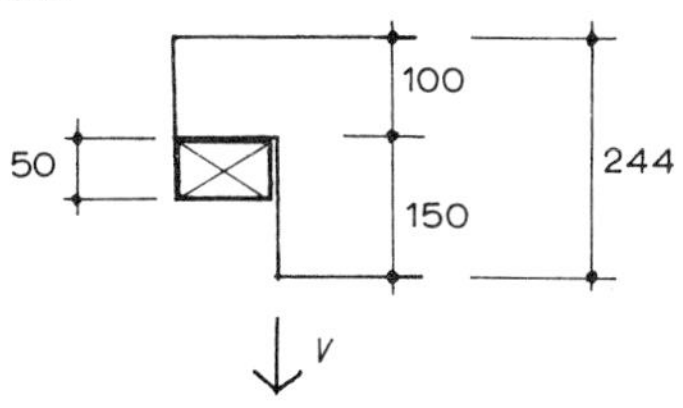

Using $\bar{V} = v_p \times d \times b \times \frac{2}{3}$ (rectangular section)

$$V = 8.75 \text{ kN}$$

$$v_p = 0.86 \text{ N/mm}^2$$

$$\bar{V} = \frac{0.86 \times 100 \times 167 \times 2}{1000 \times 3} = 9.57 \text{ kN}$$

This shear capacity must now be reduced by the factor of K_{14} to allow for the concentrated stress created by the notch cut as follows:

$$K_{14} = \frac{100}{244} = 0.409$$

Hence modified $\bar{V} = 9.57 \times 0.409 = 3.92$ kN

It can be seen that the shear capacity of the notch is considerably less than the applied shear. The simplest solution here is locally to remove the wall plate and allow the beam to sit directly on to the blockwall. This is possible because no uplift occurs at this level and also the beam is totally held in position by virtue of the floor boards connected to it.

A check on the shear capacity may now be made for an increased nib depth of 150 mm

$$K_{14} = \frac{150}{244} = 0.615$$

$$\bar{V} = \frac{0.86 \times 0.615 \times 150 \times 167 \times 2}{1000 \times 3} = 8.83 \text{ kN} > 8.75$$

This solution is therefore proven to be satisfactory.

Check bending stress

$$M = \frac{WL}{8} = 17.5 \times \frac{4.28}{8} = 9.37 \text{ kN m}$$

$$Z_x = 16.7 \times \frac{24.4^2}{6} = 1657 \text{ cm}^3$$

$$f_{apar} = \frac{9370}{1657} = 5.66 \text{ N/mm}^2$$

$$f_{ppar} = 7.30 \text{ N/mm}^2 > 5.66 \quad \text{OK}$$

Final solution

PROVIDE: 2/60 × 244 + 1/47 × 244 solid timber sections bolted together in European Whitewood to SS grade

Note: 1. Bolts to be 12 mm ϕ × 190 mm hex bolts with 50 mm × 3 mm thick washers spaced at 1.2 m c/c (4 No. required), drilled through centre of depth.
2. Notch cut depth to LH support to be a maximum of 94 mm.

Case Study 3 – Principal trimmer beam to domestic staircase

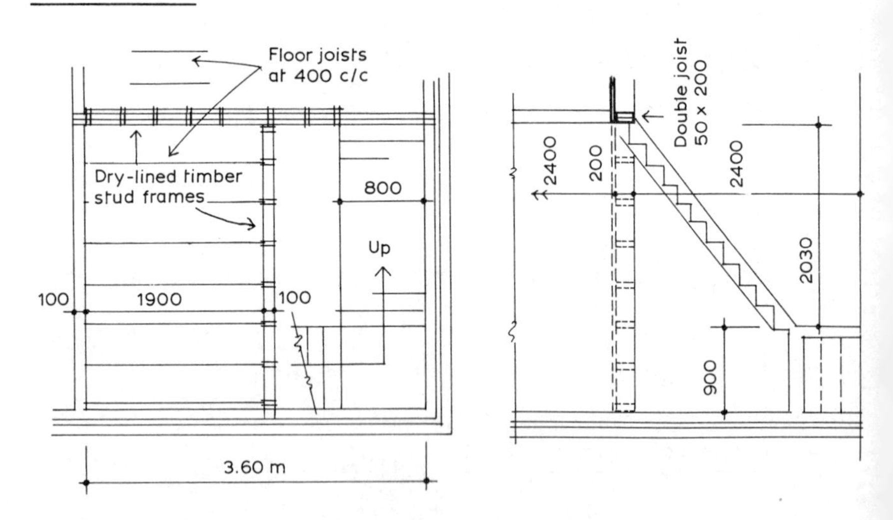

FLOOR SPECIFICATION: Domestic loading with boarded and joist floor with direction of joists as indicated. 12 mm plasterboard + skim. Dry-lined stud partition with 50 × 100 at 600 c/c and 12 mm plasterboard each side, 2.4 m high.

TIMBER: European Whitewood

STRESS GRADED: M75

TASK:
The principal trimmer beam in a house of a standard range has been rejected under the Building Regulations as being unfit for the purpose for which it is designed.

1. Determine reason for rejection.
2. Suggest modification required and justify by checking all design criteria.

Solution to Case Study 3

References

1. CP 112:Part 2:1971 'The structural use of timber'
2. CP 3:Chapter V, Part 1:1967 'Dead and imposed loading'

General loading

Floor:

Super (CP3, Chapter V, Part 1:1967)		1.50
19 mm floor boarding	0.10	
38 × 200 floor joists at 400 c/c	0.10	
12 mm plasterboard + skim	0.18	0.38
		1.88 kN/m^2

Stud partitions:

12 mm plasterboard × 2	0.26	
Framing 50 × 100 at 600 c/c	0.04	0.30 kN/m^2

Trimmer beam loading

Take as simply supported with loading as indicated:

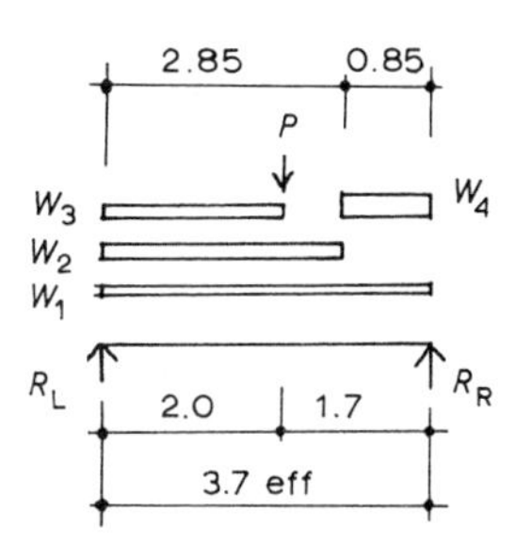

W_1 : Floor $1.88 \times 0.20 \times 3.6 = 1.36$

W_2 : Part. $0.30 \times 2.4 \times 2.8 = 2.02$

W_3 : Floor $1.88 \times 0.20 \times 1.9 = 0.72$

W_4 : Stairs $1.88 \times \frac{2.03}{2} \times 0.80 = 1.53$

5.63 kN

P : Floor $1.88 \times \frac{1.9}{2} \times \frac{2.93}{2} = 2.62$

Part. $0.30 \times 2.4 \times \frac{2.93}{2} = 1.06$

3.68 kN

Examine reason for rejection

Because we are dealing with a standard procedure of providing double joists the most likely reason for rejection is inadequate resistance to deflection. Therefore the theoretical total deflection will be examined.

As it is a check calculation then an approximation is quite in order and for this case the quickest approximation is to assume a cumulative deflection created by a UDL and a PL.

Nearest approximations would be

$$d = \frac{5WL^3}{384EI} + \frac{PL^3}{48EI}$$

where $W = W_1 + W_2 + W_3 + W_4$

Section $= 2/50 \times 200$ sawn sections

$I_x = 33.3 \times 2 = 66.6 \times 10^6 \text{ mm}^4$ $E_N = E_2 = 7870 \text{ N/mm}^2$

$$d = \frac{5 \times 5630 \times 3700^3}{384 \times 7870 \times 66.6 \times 10^6} = 7.1$$

$$\frac{3680 \times 3700^3}{48 \times 7870 \times 66.6 \times 10^6} = \frac{7.4}{14.5}$$

$$d_p = 0.003 \times 3700 = 11.1 < 14.5$$

This result indicates that the theoretical deflection is exceeded and so two joists do not have the required stiffness. By increasing the joists to 4 No. it can be seen without further calculation that the applied deflections would be halved to 7.25 mm. However because we are examining a standard design of repetitive usage a further investigation is justified in order to obtain the most economical answer. Therefore, it should be determined if 3 No. joists are stiff enough as follows:

$$I_x = 33.3 \times 3 = 99.9 \times 10^6\ \text{mm}^4\ E_N = E_3 = 8390\ \text{N/mm}^2$$

$$d = \frac{5 \times 5630 \times 3700^3}{384 \times 8390 \times 99.9 \times 10^6} = 4.43$$

$$\frac{3680 \times 3700^3}{48 \times 8390 \times 99.9 \times 10^6} = 4.63$$

$$9.06\ \text{mm} < 11.1 \qquad \text{OK}$$

Check bending stress

Determine point of zero shear

$$R_L = \frac{1.36}{2} + 2.02 \times \frac{2.3}{3.7} + 0.72 \times \frac{2.7}{3.7} + 1.53 \times \frac{0.43}{3.7} + 3.68 \times \frac{1.7}{3.7}$$

$$= 0.68 + 1.26 + 0.53 + 0.18 + 1.69$$

$$= 4.34\ \text{kN}$$

By sketching the shear force diagram to the left of the PL it can be shown that the zero shear occurs at the PL position:

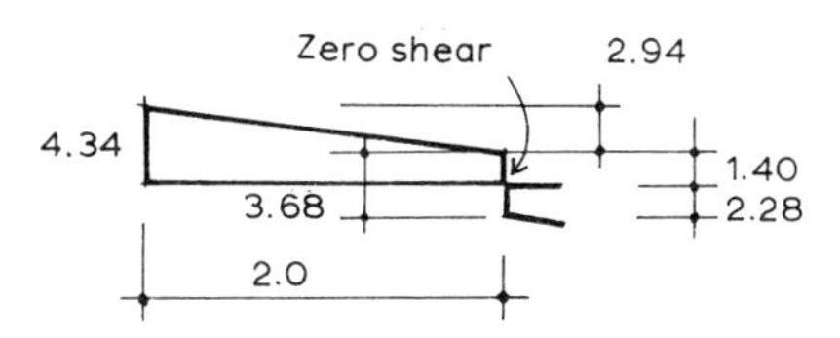

Hence the maximum bending moment is equal to the area of the shear force diagram to one side of the point of zero shear

$$\text{i.e. } M_{(\text{max.})} = \frac{4.34 + 1.40}{2} \times 2.0 = 5.74\ \text{kN m}$$

$$Z_x = 333 \times 3 = 999\ \text{cm}^3$$

$$f_{\text{apar}} = \frac{5740}{999} = 5.75\ \text{N/mm}^2$$

$$f_{\text{ppar}} = 10.0\ \text{N/mm}^2 > 5.75 \qquad \text{OK}$$

Check shear stress

To determine the maximum shear stress, it is necessary to check the magnitude of R_R and compare this with R_L

$$R_R = 5.63 + 3.68 - 4.34 = 4.97 \text{ kN} > R_L$$

The maximum shear is therefore R_R.

$$\text{Area of section} = 100 \times 3 = 300 \text{ cm}^2$$

$$v \text{ (rectangular section)} = \frac{V \times 3}{2A} = \frac{49.7 \times 3}{2 \times 300} = 0.250 \text{ N/mm}^2$$

$$v_p = 1.28 \text{ N/mm}^2 > 0.250 \qquad \text{OK}$$

Check bearing capacity

The maximum bearing capacity should be checked and compared with the maximum shear:

$$C_{pperp} = 1.8 \text{ N/mm}^2$$

$$\text{Bearing capacity} = \frac{100 \times 47 \times 3 \times 1.8}{1000} = 25.38 \text{ kN} > 4.97 \qquad \text{OK}$$

Modification required

The trimmer should be strengthened by providing an additional 50 × 200 M75 joist and securely nailing it alongside the existing double joists

Case Study 4 – Main beam supports to warehouse floor

FLOOR SPECIFICATION: Warehouse storing grain with a specified super load of 5.0 kN/m^2 and construction as indicated.

TIMBER: Douglas Fir

STRESS GRADE: SS – rough sawn section

TASK:

1. Determine minimum theoretical width (b) and relate this to nearest practicable width.

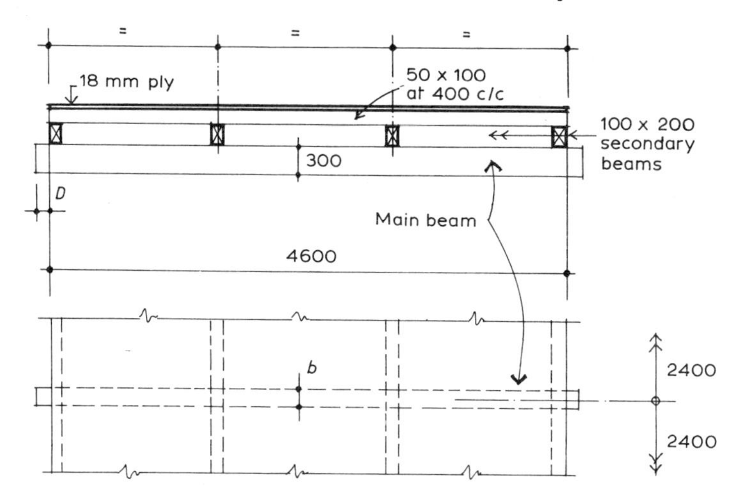

2. Determine minimum bearing (D) into wall support.
3. Check bending and shear stresses.

Solution to Case Study 4

References

1. CP 112:Part 2:1971 'The structural use of timber'
2. CP 3:Chapter V, Part 1:1967 'Dead and imposed loading'

General loading

Floor:

Super (as specified)		5.00
18 mm ply	0.10	
50 × 100 floor joists at 400 c/c	0.07	0.17
		5.17 kN/m²

Main beam loading 2.4 m c/c

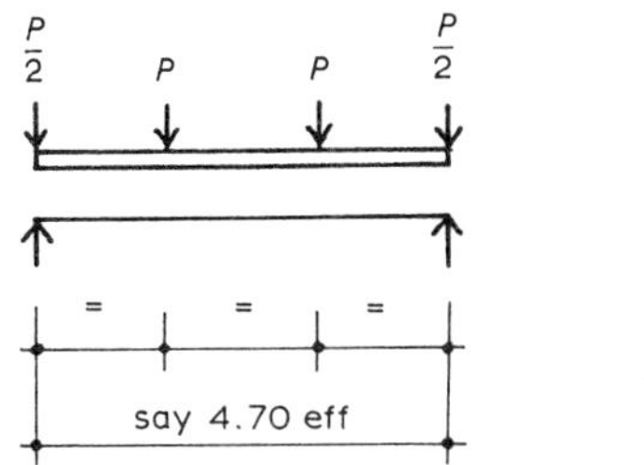

$$P: \text{Floor } 5.17 \times 2.4 \times \frac{4.7}{3} = 19.48$$

$$\text{sec. bm. } 0.102 \times 2.4 = 0.25$$

$$19.73 \text{ kN}$$

W = self weight of beam

Determine width b

The deflection will be a combination of the effect of the two equally spaced point loads coupled with the UDL (the latter will be nominal in comparison with the point loads and so will be ignored).

For two point loads equally spaced in the span,

$$d = \frac{23PL^3}{648EI_x}$$

with $E = 7300\ \mathrm{N/mm^2}$ (dry value because of grain storage)

By substituting a value for d into this equation it is possible to find the gross value of I_x and by equating I_x to the formula for a rectangle the width b may be found.

In this case we are dealing with a warehouse floor which does not have finishes which can be damaged nor does aesthetic appearance govern. Therefore, it is considered that a practical limitation of 0.0045 will be satisfactory for this structure.

$$\text{Hence } d_p = 0.0045 \times 4700 = 21.15\ \text{mm}$$

$$\text{and } I_x = \frac{23 \times 19\,730 \times 4700^3}{648 \times 7300 \times 21.15} = 471 \times 10^6\ \text{mm}^4$$

$$I_x = \frac{bd^3}{12} \text{ where } d = 300\ \text{mm}$$

$$\text{and so } b = \frac{471 \times 10^6 \times 12}{300^3} = 210\ \text{mm min.}$$

If it is possible to obtain the section then this would relate to a 250×300 solid sawn piece.

Check bearing depth (D)

$$C_{\text{pperp}} = 2.00\ \text{N/mm}^2$$

$$V = P \times 1.5 = 19.73 \times 1.5 = 29.6\ \text{kN}$$

$$\text{Area required} = \frac{296}{2.00} = 148\ \text{cm}^2$$

$$D = \frac{148}{25} = 5.92\ \text{cm} < 10\ \text{cm} \qquad \text{assumed}$$

Provide a minimum of 75 mm bearing into each supporting wall.

Check bending stress

From standard reference tables: $M = \dfrac{PL}{3}$

$$M = 19.73 \times \frac{4.7}{3} = 30.91 \text{ kN m}$$

$$Z_x = 25 \times \frac{30^2}{6} = 3750 \text{ cm}^3$$

$$f_{\text{apar}} = \frac{30\,910}{3750} = 8.25 \text{ N/mm}^2$$

$$f_{\text{ppar}} = 9.3 \text{ N/mm}^2 > 8.25 \qquad \text{OK}$$

Check shear capacity

$$V = 29.6 \text{ kN (as before)}$$

$$v_{\text{p}} = 0.96 \text{ N/mm}^2$$

$$\bar{V} = 0.250 \times 300 \times 0.96 \times \frac{2}{3} = 48 \text{ kN} > 29.6 \qquad \text{OK}$$

Final section

(a) 250 × 300 sawn SS grade in Douglas Fir-Larch
(b) 300 deep by any combination of laminate thicknesses bolted together to give a 250 mm o/a thickness. Minimum 4 No. 16 mm ϕ bolts spaced 600 mm from wall face and equally between, with 75 mm square × 4 mm thick washers each side.

PROP COLUMN SUBJECT TO AXIAL LOAD ONLY

Case Study 5

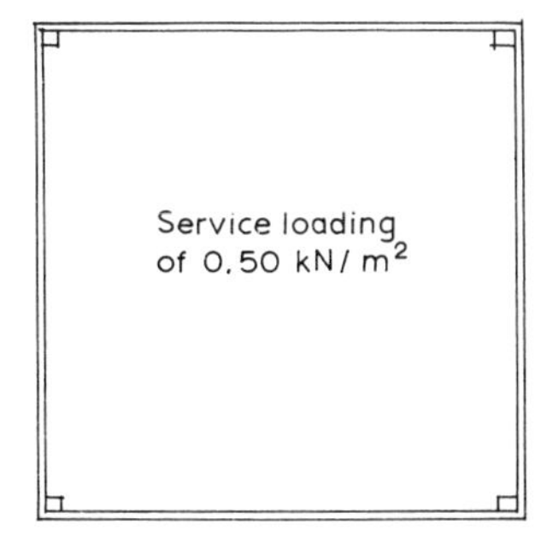

ROOF SPECIFICATION:
12 mm chippings
3 layers roofing felt
19 mm t and g boarding
50 × 250 at 400 c/c
60 mm insulation
2 layers 12.7 mm plasterboard
1 coat skim

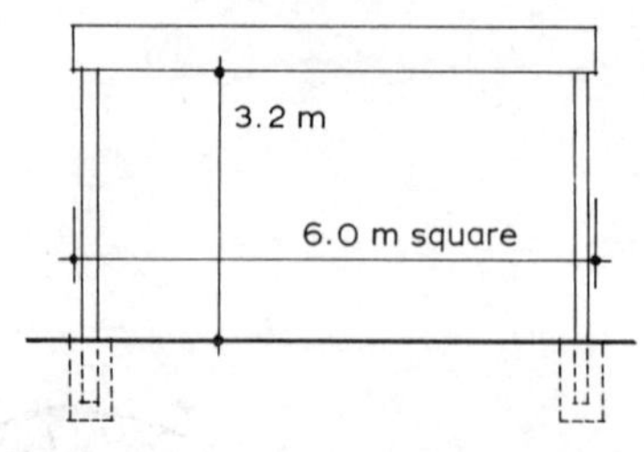

TIMBER:
Canadian Spruce stress graded to 40 visual with stress levels to Table 3 of CP 112:Part 2:1971.

TASK:
Ignoring the self weight of the edge beams determine the minimum size of corner posts required to carry the roof construction.

It may be assumed that the posts will be square in section.

Solution to Case Study 5

References

1. CP 112:Part 2:1971 'The structural use of timber'
2. CP 3, Chapter V, Part 2:1967 'Loading'
3. BS 648:1964 'Weights of building materials'

General loading

Super		0.75
Dead:	12 mm chippings	0.20
	3 layers felt	0.11
	19 mm t and g boarding	0.11
	50 × 250 at 400 c/c	0.16
	60 mm insulation	0.03
	2 layers 12.7 mm plasterboard	0.26
	1 coat skim	0.05
		1.67 kN/m^2
	Service load	0.50
		2.17 kN/m^2

Line loading diagram

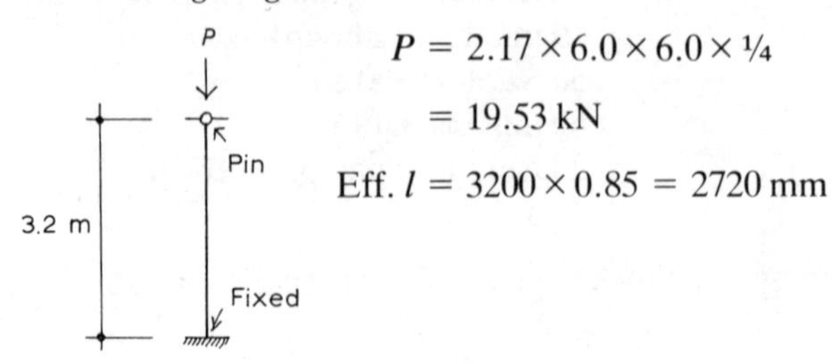

$$P = 2.17 \times 6.0 \times 6.0 \times 1/4$$

$$= 19.53 \text{ kN}$$

$$\text{Eff. } l = 3200 \times 0.85 = 2720 \text{ mm}$$

Try 75 mm square

Properties:

$$A = 7.5^2 = 56.25 \text{ cm}^2 \qquad I = 7.5 \times \frac{7.5^3}{12} = 263.7 \text{ cm}^4$$

$$r = \sqrt{\frac{263.7}{56.25}} = 2.17 \text{ cm} \qquad \frac{l}{r} = \frac{272}{2.17} = 126$$

$K_{18} = 0.39$ (med. term)

From Table 3 $C_{gpar} = 3.1 \text{ N/mm}^2$

hence $C_{ppar} = 3.1 \times 0.39 = 1.209 \text{ N/mm}^2$

$$C_{apar} = \frac{195.3}{56.25} = 3.472 \text{ N/mm}^2 > 1.209 \qquad \text{No!}$$

Try 100 mm square

$$A = 100 \text{ cm}^2 \qquad I = 10 \times \frac{10^3}{12} = 833.3 \text{ cm}^4$$

$$r = \sqrt{\frac{833.3}{100}} = 2.88 \text{ cm} \qquad \frac{l}{r} = \frac{272}{2.88} = 94$$

$K_{18} = 0.64$

$$C_{pper} = 3.1 \times 0.64 = 1.984 \text{ N/mm}^2$$

$$C_{apar} = \frac{195.3}{100} = 1.953 \text{ N/mm}^2 < 1.984 \qquad \text{OK}$$

Selected section

PROVIDE: 100 mm square sawn section in Canadian Spruce to 40 visual grade

Case Study 6

The floor loading is to be taken as 3.33 kN/m² and the self weight of the beam is 0.5 kN/m.

TASK:
Assuming that the column size is fixed beyond alteration determine the applied axial stress and from this specify the timber species and grade to be used.

The column is to be built into a timber frame wall which is 72 mm wide.

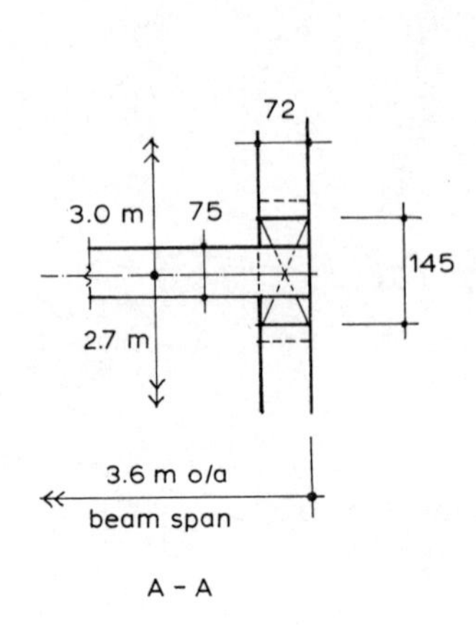

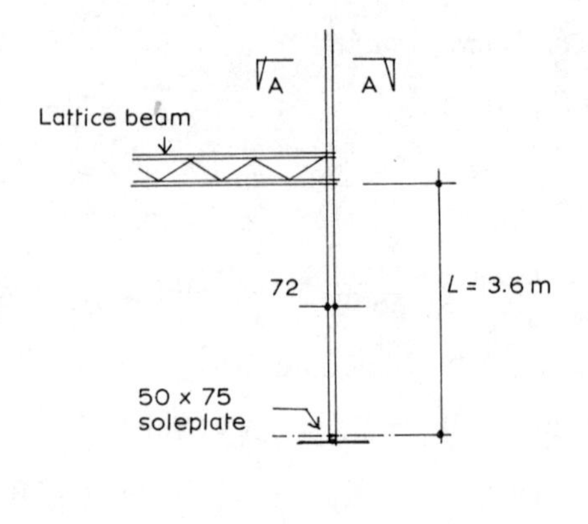

Solution to Case Study 6

Reference

CP 112:Part 2:1971 'The structural use of timber'

Column loading

Beam reaction $3.3 \times \frac{5.7}{2} \times \frac{3.6}{2} = 17.1$

Self weight beams $0.5 \times 1.8 = 0.90$

Col. self weight say $= 0.30$

18.3 kN

Applied stress

$$C_{apar} = \frac{183}{7.2 \times 14.5} = 1.75 \text{ N/mm}^2$$

Column properties

$$A = 7.2 \times 14.5 = 104.4 \text{ cm}^2 \quad I_y = 14.5 \times \frac{7.2^3}{12} = 451 \text{ cm}^4$$

$$r_y = \sqrt{\frac{451}{104.4}} = 2.08 \text{ cm}$$

Taking the column to be pinned both at the top and at the bottom then:

$$l = L = 360 \text{ cm} \qquad \frac{l}{r} = \frac{360}{2.08} = 173$$

limiting l/r is 180 therefore 173 is acceptable.

Assuming that a soft wood will be satisfactory, then factor K_{18} applies and in this case must be taken as long term because the loading is via a floor.

From Table 15 $K_{18} = 0.217$

Now equating the permissible stress to the applied stress we have:

$$C_{\text{gpar}} \times 0.217 = 1.75$$

$$\text{giving } C_{\text{gpar}} = \frac{1.75}{0.217} = 8.07 \text{ N/mm}^2$$

Compare with a hardwood species and re-examine the modification factor using K_{19} as follows:

Try Keruing with $E_{\text{min.}} = 9300 \text{ N/mm}^2$

$$\text{limiting } \frac{l}{r} \leq \sqrt{\left(\frac{11.46 \times E}{C_{\text{gpar}}}\right)}$$

as the stress is high usc a high grade
i.e. 75 grade giving $C_{\text{gpar}} = 13.1 \text{ N/mm}^2$

$$\text{Hence limit} = \sqrt{\left(\frac{11.46 \times 9300}{13.1}\right)} = 90.2 < 173$$

$$\text{Therefore } 5.73 \frac{E}{C_{\text{gpar}}}\left(\frac{r}{l}\right)^2 \text{ applies}$$

$$K_{19} = \frac{5.73 \times 9300}{13.1} \times \frac{2.08^2}{360^2} = 0.136$$

$$C_{\text{ppar}} = 13.1 \times 0.136 = 1.78 \text{ N/mm}^2 > 1.75 \qquad \text{OK}$$

Selection

PROVIDE:
Any one of following species with grade as indicated:

Western Hemlock (commercial)	75 visual grade
Pitch Pine	65 visual grade
Keruing	75 visual grade

COLUMN SUBJECT TO BENDING AND AXIAL LOAD

Case Study 7

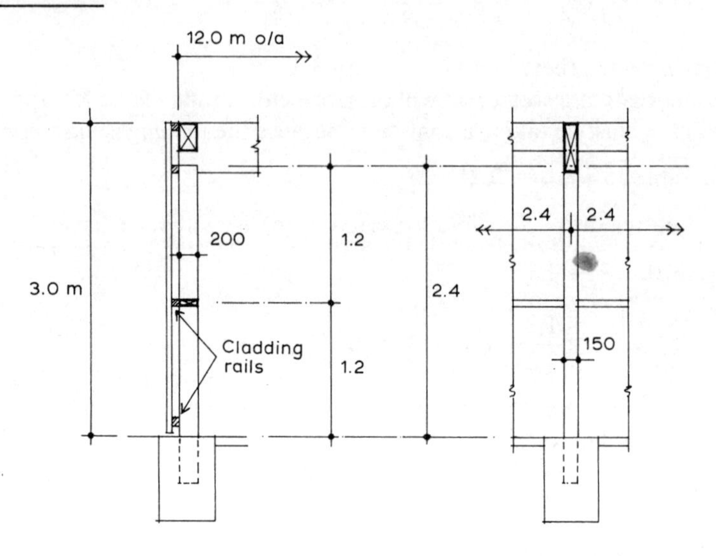

SPECIFICATION:
Roof loading to be: Super 0.75 kN/m^2, Dead 0.85 kN/m^2
Timber to be European Redwood or Whitewood stressed graded to SS standard.

TASK:
The column shown is part of an industrial building and deflection of the column is not considered critical. The building is to be erected in the suburbs of Edinburgh. Check that the column size indicated satisfies the requirements of Clause 3.14.3 of CP 112:Part 2:1971.

Solution to Case Study 7

References

1. CP 112:Part 2:1971 'The structural use of timber'
2. CP 3, Chapter V, Part 2:1972 'Wind loading'

General loading

As given in the specification: Super = 0.75
Dead = 0.85

1.60 kN/m^2

Wind loading (q)

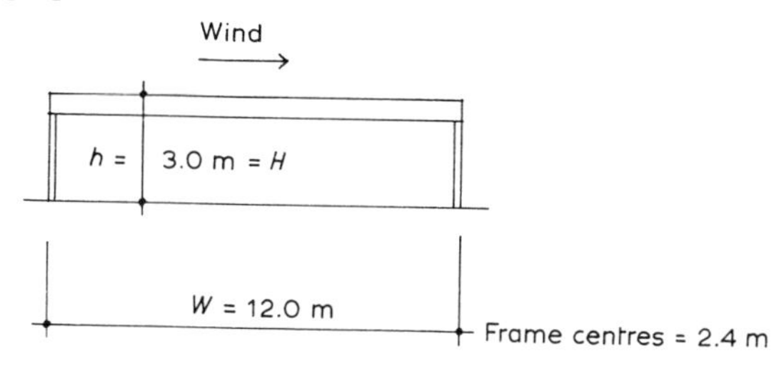

Location is Edinburgh; from Table 1 of CP3, Chapter V basic wind speed $V = 50$ m/s

Take $S_1 = S_3 = 1.0$

From Table 3, with $H = 3.0$ m and using Class B, Group (3) $S_2 = 0.600$

Giving $V_s = 50 \times 1.0 \times 0.600 \times 1.0 = 30$ m/s

$q = 0.613V_s^2 = 0.613 \times 30^2 = 551$ N/m^2

Pressure coefficients and (q_d)

Take the total pressure coefficients for the walls $C_{pe} + C_{pi} = 1.0$

$q_d = q = 551$ N/m^2 (walls)

For the roof from Table 8 with $\alpha = 0°$ and h/w by inspection $<\frac{1}{2}$ then average uplift

$$\text{coefficients } C_{pe} \text{ av.} = \frac{-0.8 - 0.4}{2}$$

$$= -0.60$$

take $C_{pi} = +0.20$ giving a total uplift coefficient of

$C_{pe} + C_{pi} = -0.60 - (+0.20) = -0.80$

Hence $q_d = 551 \times -0.80 = -441$ N/m^2 (roof)

Check for uplift

Roof:	unit dead load	$= +0.85$
	unit wind uplift	$= -0.44$
	net unit loading	$+0.41$ kN/m^2

WIND UPLIFT DOES NOT OCCUR

Line loading diagram

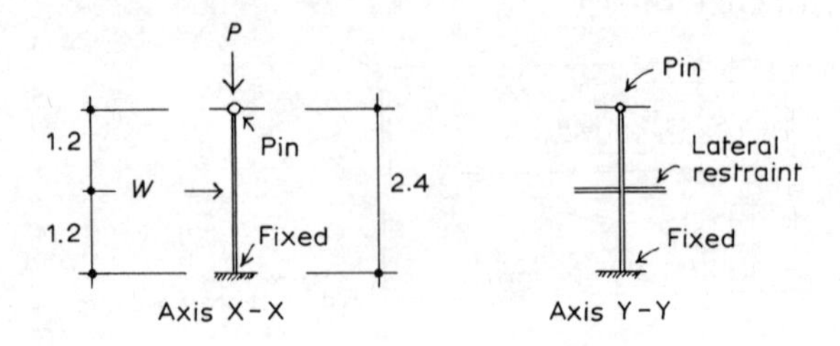

Effective lengths

$$l_x = 0.85 \times 2400 = 2040 \text{ mm}$$

$$l_y = 2400 \times 0.85 \times \tfrac{1}{2} = 1020 \text{ mm}$$

Loading and wind moment

$$P\text{: Dead + Wind } 0.41 \times 2.4 \times \frac{12}{2} = 5.91$$

$$\text{Super } 0.75 \times 2.4 \times \frac{12}{2} = \underline{10.80}$$

$$16.71 \text{ kN}$$

$$W\text{: Wind } 0.551 \times 2.4 \times 1.2 = 1.59 \text{ kN}$$

$$\text{Wind moment} = 1.59 \times \frac{2.4}{4} = 0.954 \text{ kN m}$$

Section

Post size × 200 × 150 actual

Properties and permissible stresses

$$A = 300 \text{ cm}^2 \quad Z_x = 1000 \text{ cm}^3 \quad r_x = 57.7 \text{ mm} \quad r_y = 43.3 \text{ mm}$$

$$\frac{l_x}{r_x} = \frac{2040}{57.7} = 35.4 \quad \frac{l_y}{r_y} = \frac{1020}{43.3} = 24$$

$$\text{limiting } \frac{l}{r} = 35.4$$

From Table 15

$$K_{18} = 1.15 \text{ (med. term)}$$

$$C_{gpar} \text{ (SS grade)} = 8.0\,\text{N/mm}^2$$

$$C_{ppar} = 8.0 \times 1.15 = 9.2\,\text{N/mm}^2$$

$$f_{gpar} = 7.3\,\text{N/mm}^2$$

$$f_{ppar} = 7.3 \times 1.50 = 10.95\,\text{N/mm}^2$$

Applied stresses

$$C_{apar} = \frac{P}{A} = \frac{167.1}{300} = 0.58\,\text{N/mm}^2$$

$$f_{apar} = \frac{M}{Z_x} = \frac{954}{1000} = 0.96\,\text{N/mm}^2$$

Summation of stresses

Clause 3.14.3:

$$\frac{f_{apar}}{f_{ppar}} + \frac{C_{apar}}{C_{ppar}} \not> 0.9$$

$$\frac{0.96}{10.95} + \frac{5.57}{9.2}$$

$$0.088 + 0.063 = 0.151 < 0.90 \qquad \text{OK}$$

Note: In these calculations the dead load of the supporting framework has not been included but it can be seen that neither the applied compressive stress nor the summation factor would be exceeded by the inclusion of this small increase in load.

Case Study 8

TASK:
Given the cross-section through the structural framework of a two-storey timber frame house (below), determine the suitability of the standard stud to the lower external walls using:

1. 38 × 89 CLS constructional grade timber with grade stresses to Table 65 of CP 112, Part 2:1971.

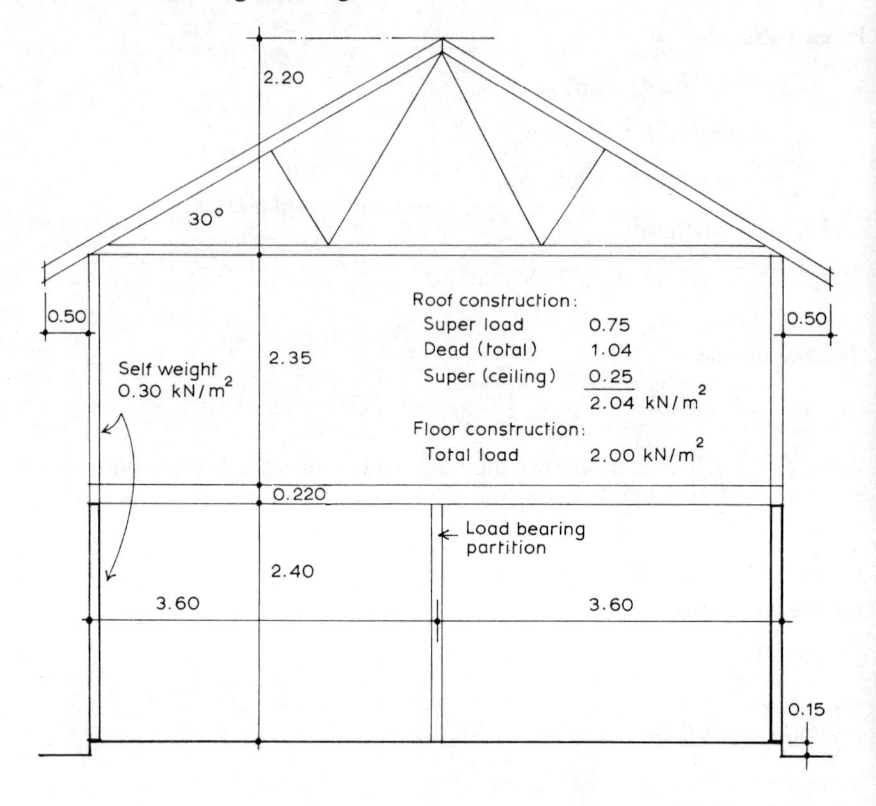

2. Basic Wind speed = 46 m/s, $S_1 = S_3 = 1.0$ and Group 3 for assessing S_2.
3. All roof trusses, studs and floor joists line through on a 600 mm grid.
4. With the length of the house equal to 10.0 m check for its stability and calculate the total racking forces induced in the building.

Solution to Case Study 8

References

1. CP 112:Part 2:1971 'The structural use of timber'
2. CP 3, Chapter V, Part 2:1972 'Wind loading'
3. Specification given on sketch details.

General loading

As given in the specification:
Roof = 2.04 kN/m^2 total
Floor = 2.0 kN/m^2 total
Walls = 0.30 kN/m^2

Wind loading

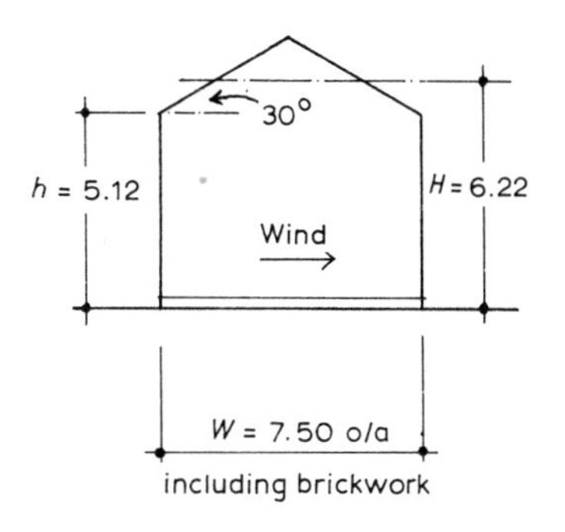

Given:
$V = 46$ m/s
$S_1 = S_3 = 1.0$
All members at 600 c/c

Note: CP 3, Chapter V, allows the building height to be broken down into component parts when assessing the wind pressure; therefore, it is reasonable to take the average overall height for (H) as indicated.

From Table 3 with $H = 6.22$ m and using Class B Group (3) $S_2 = 0.672$
Giving $V_s = 46 \times 1.0 \times 0.672 \times 1.0 = 31$ m/s
$q = 0.613\, V_s^2 = 0.613 \times 31^2 = 589$ N/m^2

Pressure coefficients and (q_d)

Take the total pressure coefficients for the walls $C_{pe} + C_{pi} = 1.0$

$$q_d = q = 589 \text{ N/m}^2 \text{ (walls)}$$

For the roof from Table 8 with $\alpha = 0°$

$$\text{and } \frac{h}{w} = \frac{5.12}{7.50} = 0.682 \left[\frac{1}{2} < \frac{h}{w} \leq \frac{3}{2}\right]$$

$$\text{Uplift coefficients } C_{pe} \text{ av.} = \frac{-0.20 - 0.50}{2 \times \cos 30°}$$

$$= -0.404 \text{ on plan}$$

take $C_{pi} = \dfrac{+0.20}{\cos 30°}$ giving a total uplift coefficient of

$$C_{pe} + C_{pi} = -0.404 - (+0.23)$$

$$= -0.634 \text{ on plan}$$

Hence $q_d = 589 \times -0.634 = -374$ N/m^2 (roof – on plan)

Check for uplift

Roof: unit dead load $= +1.040$
unit wind load $= -0.374$

net unit loading $= +0.666\ \text{kN/m}^2$

WIND UPLIFT DOES NOT OCCUR

Line loading diagram and eff. length

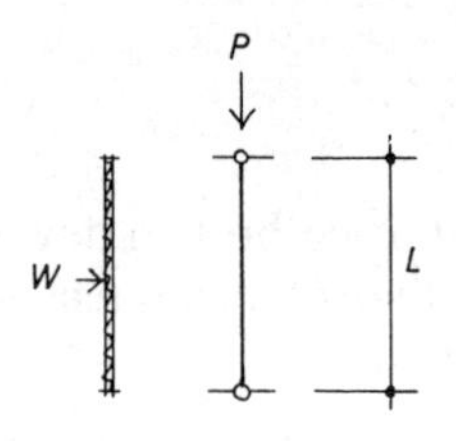

With the degree of restraint afforded by the fixing to the sole plate plus the fixing of the plywood through the bottom plate and head binder, it is reasonable to assume a reduction coefficient of 0.85 for l_x

Hence $L = 2400 - (3 \times 38) = 2286$
and $l_x = 0.85 \times 2286 = 1943$ mm

Loading and wind moment

P: Roof Dead + Wind + Super	$(0.75 + 0.666) \times 0.60 \times 3.60$	$= 3.06$
Ceiling super	$0.250 \times 0.60 \times 3.60$	$= 0.54$
Wall 1st to roof	$0.30 \times 0.60 \times 2.35$	$= 0.42$
Floor total	$2.00 \times 0.60 \times 1.80$	$= 2.16$
Wall Grd. to 1st	$0.30 \times 0.60 \times 2.40$	$= 0.43$
	say	$= 6.61$ kN

W: Wind $0.589 \times 0.60 \times 2.40 = 0.85$ kN

$$\text{Wind moment} = 0.85 \times \frac{2.286}{8} = 0.243\ \text{kN m}$$

Section properties and permissible stresses

Stud size $= 38 \times 89$ actual

$$A = 3.8 \times 8.9 = 33.82\ \text{cm}^2$$

$$Z_x = 3.8 \times \frac{8.9^2}{6} = 50.16\ \text{cm}^3$$

$$I_x = 3.8 \times \frac{8.9^3}{12} = 223.2\ \text{cm}^4$$

$$r_x = \sqrt{\frac{223.2}{33.82}} = 2.57\ \text{cm}$$

It may be safely assumed that the r_y axis is fully restrained by the plywood sheathing.

$$\frac{l_x}{r_x} = \frac{1943}{25.7} = 75.6$$

From Table 15 $K_{18} = 0.838$ (med. term)

$$C_{\text{gpar}}\ (\text{Table 65}) = 6.55\ \text{N/mm}^2$$

$$C_{\text{ppar}} = 6.55 \times 0.838 \times 1.1 = 6.04\ \text{N/mm}^2$$

$$f_{\text{gpar}} = 8.27\ \text{N/mm}^2$$

$$f_{\text{ppar}} = 8.27 \times 1.50 \times 1.1 = 13.64\ \text{N/mm}^2$$

Applied stresses

$$C_{\text{apar}} = \frac{P}{A} = \frac{66.1}{33.82} = 1.95\ \text{N/mm}^2$$

$$f_{\text{apar}} = \frac{M}{Z_x} = \frac{243}{50.16} = 4.84\ \text{N/mm}^2$$

Summation of stresses

Clause 3.14.3:

$$\frac{f_{\text{apar}}}{f_{\text{ppar}}} + \frac{C_{\text{apar}}}{C_{\text{ppar}}} \ngtr 0.90$$

$$\frac{4.84}{13.64} + \frac{1.95}{6.04}$$

$$0.355 + 0.323 = 0.678 < 0.90 \qquad \text{OK}$$

Notes:

1. As CLS timber tends to vary in species it may be necessary to check the permissible design stresses with the importer or exporter.
2. In some cases, for studs at a level below the upper floor, dead loads may be high. If so a check should be made excluding the roof super and taking K_{18} for its long-term value, i.e. total dead + wind.

Check for overturning

Use pressure coefficients, because the building contains a pitched roof and take for $C_{pi} = \pm 0.3$

Wall coefficients

From Table 7 of CP 3, Chapter V for

$$\frac{h}{w} = \frac{4.97}{7.20} = 0.69 \qquad \text{Use } 2 < \frac{h}{w} \leq \frac{3}{2}$$

$$\frac{l}{w} = \frac{10.0}{7.2} = 1.39 \qquad \text{Use } 1 < \frac{l}{w} \leq \frac{3}{2}$$

	A		*B*
For $\alpha°$: C_{pe} =	+0.7	*B* =	−0.25
C_{pi} =	−0.3	*B* =	−0.30
Algebraic $C_{pe} + C_{pi}$ =	+1.0		−0.05
=	0.95 across building		

Roof coefficients

From Table 8 of CP 3, Chapter V for

$$\frac{h}{w} = 0.69 \text{ and roof angle } = 30°$$

Use $\frac{1}{2} < \frac{h}{w} \leq \frac{3}{2}$ giving:

For $\alpha°$:	EF	GH
C_{pe} =	−0.2	−0.5
C_{pi} =	+0.3	+0.3
Algebraic $C_{pe} + C_{pi}$ =	−0.5	−0.8

acting normal to roof slope

$$\text{On plan} = \frac{-0.5}{\cos 30°} = -0.577 \text{ for EF}$$

$$\text{and} \quad \frac{-0.8}{\cos 30°} = -0.924 \text{ for GH}$$

Loading diagram

Wind + dead loads (per metre length)

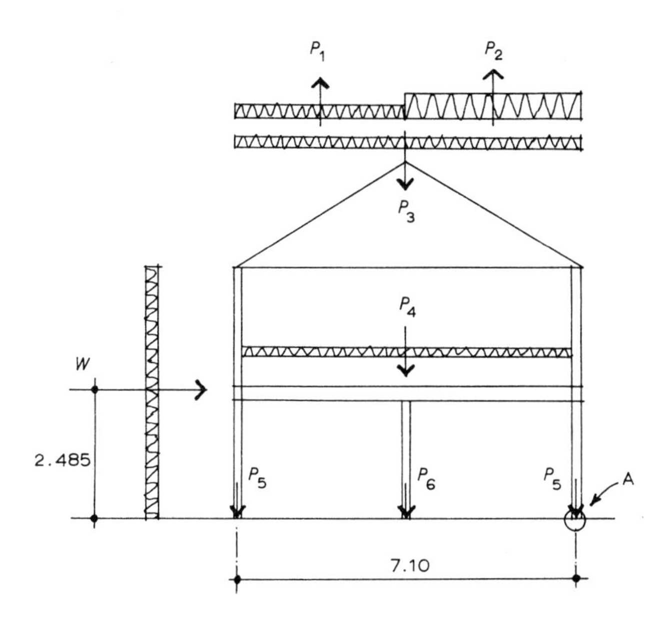

Applied loads

$W = 0.589 \times 4.97 \quad \times 0.95 \qquad = \quad 2.78\ \text{kN}$

$P_1 = 0.589 \times 0.577 \times \dfrac{7.10}{2} \qquad = -1.21\ \text{kN}$

$P_2 = 0.589 \times 0.924 \times \dfrac{7.10}{2} \qquad = -1.93\ \text{kN}$

$P_3 = (2.04 - 0.75 - 0.25) \times 7.20 \qquad = \quad 7.48$

overhang say $0.80 \times 1.0 \qquad = \quad 0.80$

$8.28\ \text{kN}$

$P_4 = (2.0 - 1.50) \times 7.0 \qquad = \quad 3.50\ \text{kN}$

$P_5 = 0.30 \times 4.97 \qquad = \quad 1.49\ \text{kN}$

$P_6 = 0.30 \times 2.40 \qquad = \quad 0.72\ \text{kN}$

Overturning moment

Taking moments about A we have:

$$\text{Wind moment} = (2.78 \times 2.485) + (1.21 \times 7.1 \times 0.75) + (1.93 \times 7.1 \times 0.25)$$

$$= 16.78 \text{ kN m}$$

Resisting moment

Taking moments about A:

Resisting moment

$$= \left(8.28 \times \frac{7.1}{2}\right) + \left(3.5 \times \frac{7.1}{2}\right) + \left(1.49 \times 7.1\right) + \left(0.72 \times \frac{7.1}{2}\right)$$

$$= 54.95 \text{ kN m}$$

Factor of safety against overturning

$$= \frac{54.95}{16.78} = 3.27 \text{ which is satisfactory}$$

Check for sliding

Total net dead load $= \Sigma(P_1 \text{ through to } P_6)$

$= 12.34$ kN

Total horizontal wind load

$$= 0.589 \times 0.95 \times (4.97 + 2.2) = 4.02 \text{ kN}$$

Factor of safety against sliding using a coefficient of friction of say 0.4

$$= \frac{12.34 \times 0.4}{4.02} = 1.23$$

This would normally be considered to be too low but it should be noted that the calculation does not take into account the weight of the external cladding material. During the construction stage the author would look to obtain a minimum factor of 1.20 (bearing in mind that with roof tiling on, fixing would in any case have been applied at soleplate level) whilst for the final construction a factor of 2.0 would be required. If, for example, this building were to be clad externally by brickwork then allowing for 50% elevational openings the increased dead load for 100 mm brickwork would be:

Brickwork dead $= 2.0 \times 4.97 = 9.94$ kN

$$\text{Increased net dead load} = 12.34 + 9.94$$
$$= 22.28 \text{ kN}$$

$$\text{Factor of safety} = \frac{22.28 \times 0.4}{4.02} = 2.22$$

which is satisfactory

Total racking forces acting on building

(a) Perpendicular to length of 10.0 m

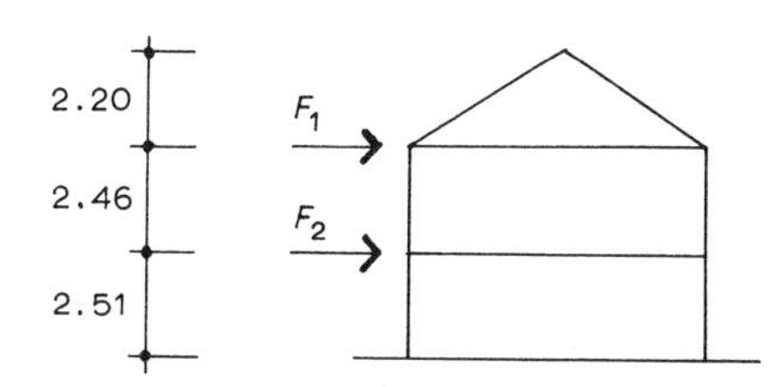

The total racking forces are F_1 and F_2 acting respectively at roof and floor diaphragm levels.

$$F_1 = 0.589 \times \left(2.2 + \frac{2.46}{2}\right) \times 10 = 20.20 \text{ kN}$$

$$F_2 = 0.589 \times \left(2.2 + 2.46 + \frac{2.51}{2}\right) \times 10 = 34.80 \text{ kN}$$

(b) Perpendicular to width of 7.2 m

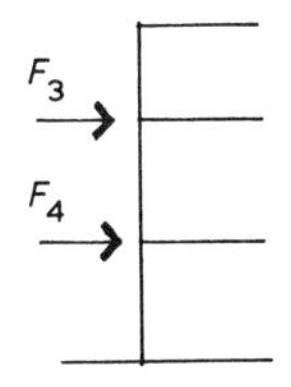

In this case F_3 and F_4 are the total racking loads to be found.

$$F_3 = 0.589 \times (2.2 \times 3.6 + 1.23 \times 7.2) = 9.88 \text{ kN}$$

$$F_4 = F_3 + 0.589 \times 7.2 \times (1.23 + 1.25) = 20.40 \text{ kN}$$

Total forces induced into perimeter walls

(a) Each long wall

$$\text{upper floor} = \frac{F_1}{2} = \frac{20.20}{2} = 10.10\ \text{kN}$$

$$\text{ground floor} = \frac{F_2}{2} = \frac{34.80}{2} = 17.90\ \text{kN}$$

(b) Each short wall

$$\text{upper floor} = \frac{F_3}{2} = \frac{9.88}{2} = 4.94\ \text{kN}$$

$$\text{ground floor} = \frac{F_4}{2} = \frac{20.40}{2} = 10.20\ \text{kN}$$

Final comment

Each wall would need to be examined for the specified racking resistance measured against the total dead load carried. The design procedure would follow the parameters explained in Example 11.3. Should there prove to be insufficient length internal racking panels would be chosen to supplement the building's total resistance.

Appendix A: Simply supported beam information

The following tables are used on numerous occasions to determine the maximum moments, shears and deflections in simply supported beams where various patterns of loading commonly feature in design.

They are a simple, quick and easy means of shortening the time taken to solve the problem and it is normally sufficient, when preparing the structural calculations, to give a sketch of the loading pattern and make the statement: 'from standard reference tables for the loading pattern indicated the following moments, shears and deflections are determined'.

The last two tables, which indicate a partial uniformly distributed load and a concentrated load, give values of K_b which when multiplied by the actual load occurring provide an equivalent total uniformly distributed load by which the deflection may be related to $5WL^3/EI$. Again, their use offers a short cut which should be acceptable to the majority, and they save time in an area of design where these two patterns are extremely common and where it is sufficiently accurate to summate all deflection values obtained to a maximum centre span value.

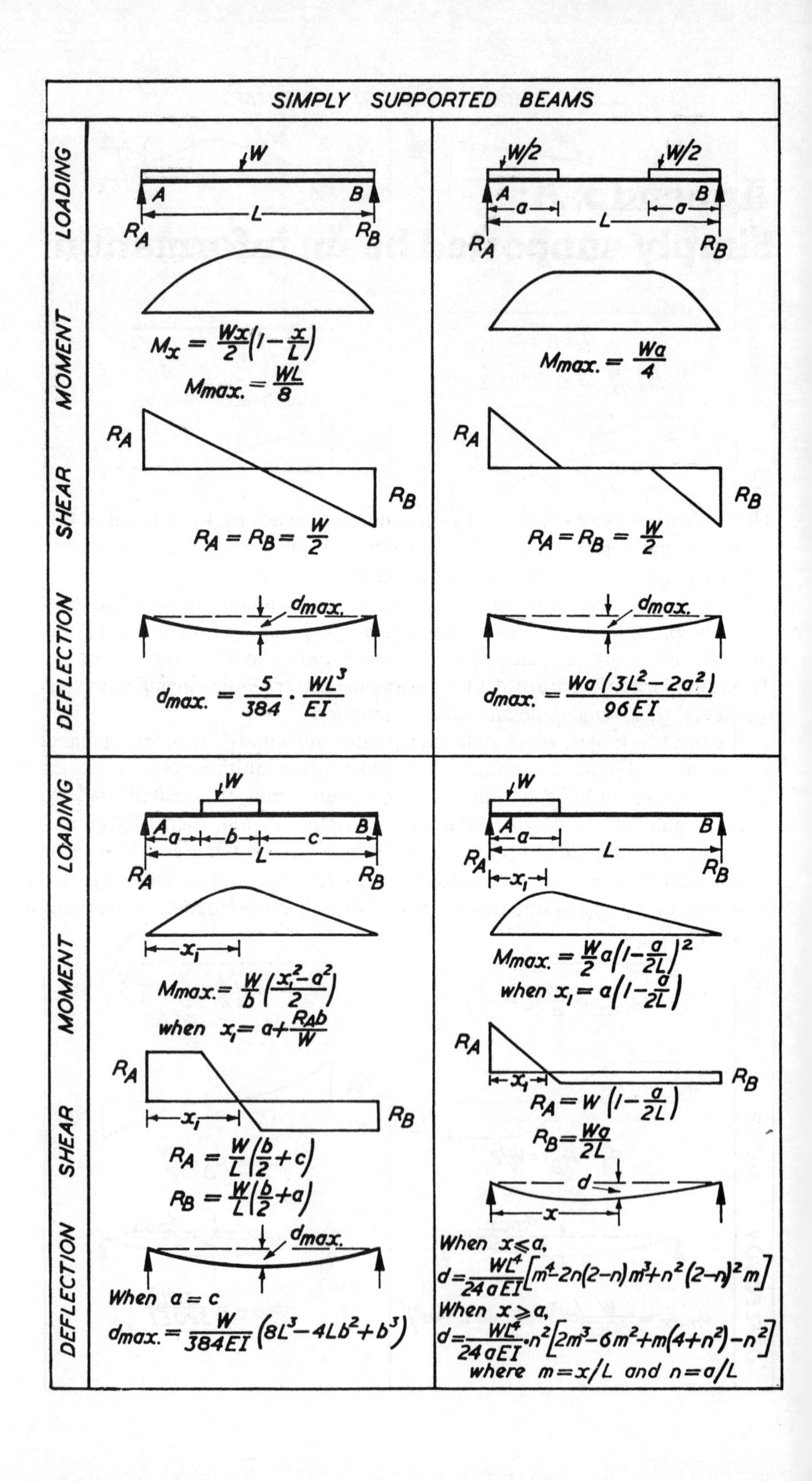

SIMPLY SUPPORTED BEAMS
LOADING
MOMENT
SHEAR
DEFLECTION
W
A
B
L
R_A
R_B
$M_x = \frac{Wx}{2}\left(1-\frac{x}{L}\right)$
$M_{max.} = \frac{WL}{8}$
$R_A = R_B = \frac{W}{2}$
$d_{max.}$
$d_{max.} = \frac{5}{384}\cdot\frac{WL^3}{EI}$
W/2
W/2
a
a
$M_{max.} = \frac{Wa}{4}$
$R_A = R_B = \frac{W}{2}$
$d_{max.} = \frac{Wa(3L^2-2a^2)}{96EI}$
W
a
b
c
x_1
$M_{max.} = \frac{W}{b}\left(\frac{x_1^2-a^2}{2}\right)$
when $x_1 = a + \frac{R_A b}{W}$
$R_A = \frac{W}{L}\left(\frac{b}{2}+c\right)$
$R_B = \frac{W}{L}\left(\frac{b}{2}+a\right)$
When $a = c$
$d_{max.} = \frac{W}{384EI}(8L^3-4Lb^2+b^3)$
W
a
x_1
$M_{max.} = \frac{W}{2}a\left(1-\frac{a}{2L}\right)^2$
when $x_1 = a\left(1-\frac{a}{2L}\right)$
$R_A = W\left(1-\frac{a}{2L}\right)$
$R_B = \frac{Wa}{2L}$
d
x
When $x \leqslant a$,
$d = \frac{WL^4}{24aEI}\left[m^4 - 2n(2-n)m^3 + n^2(2-n)^2 m\right]$
When $x \geqslant a$,
$d = \frac{WL^4}{24aEI}\cdot n^2\left[2m^3 - 6m^2 + m(4+n^2) - n^2\right]$
where $m = x/L$ and $n = a/L$

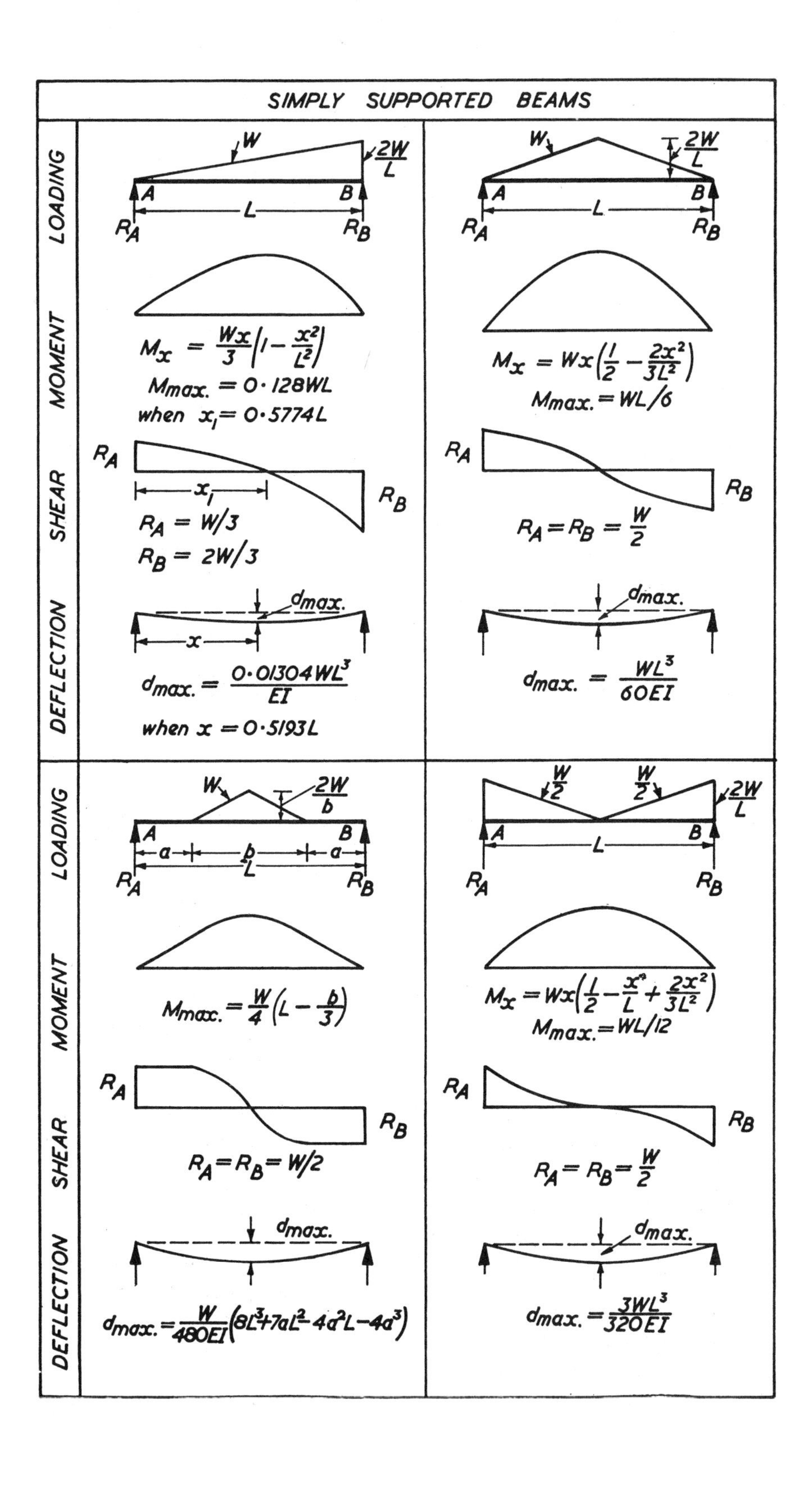
SIMPLY SUPPORTED BEAMS
LOADING
MOMENT
SHEAR
DEFLECTION
W
2W/L
A
B
L
R_A
R_B
M_x = (Wx/3)(1 − x²/L²)
M_max. = 0·128WL
when x₁ = 0·5774L
x₁
R_A = W/3
R_B = 2W/3
d_max.
x
d_max. = 0·01304WL³/EI
when x = 0·5193L
M_x = Wx(1/2 − 2x²/3L²)
M_max. = WL/6
R_A = R_B = W/2
d_max. = WL³/60EI
2W/b
a
b
M_max. = (W/4)(L − b/3)
R_A = R_B = W/2
d_max. = (W/480EI)(8L³+7aL²−4a²L−4a³)
W/2
M_x = Wx(1/2 − x/L + 2x²/3L²)
M_max. = WL/12
d_max. = 3WL³/320EI

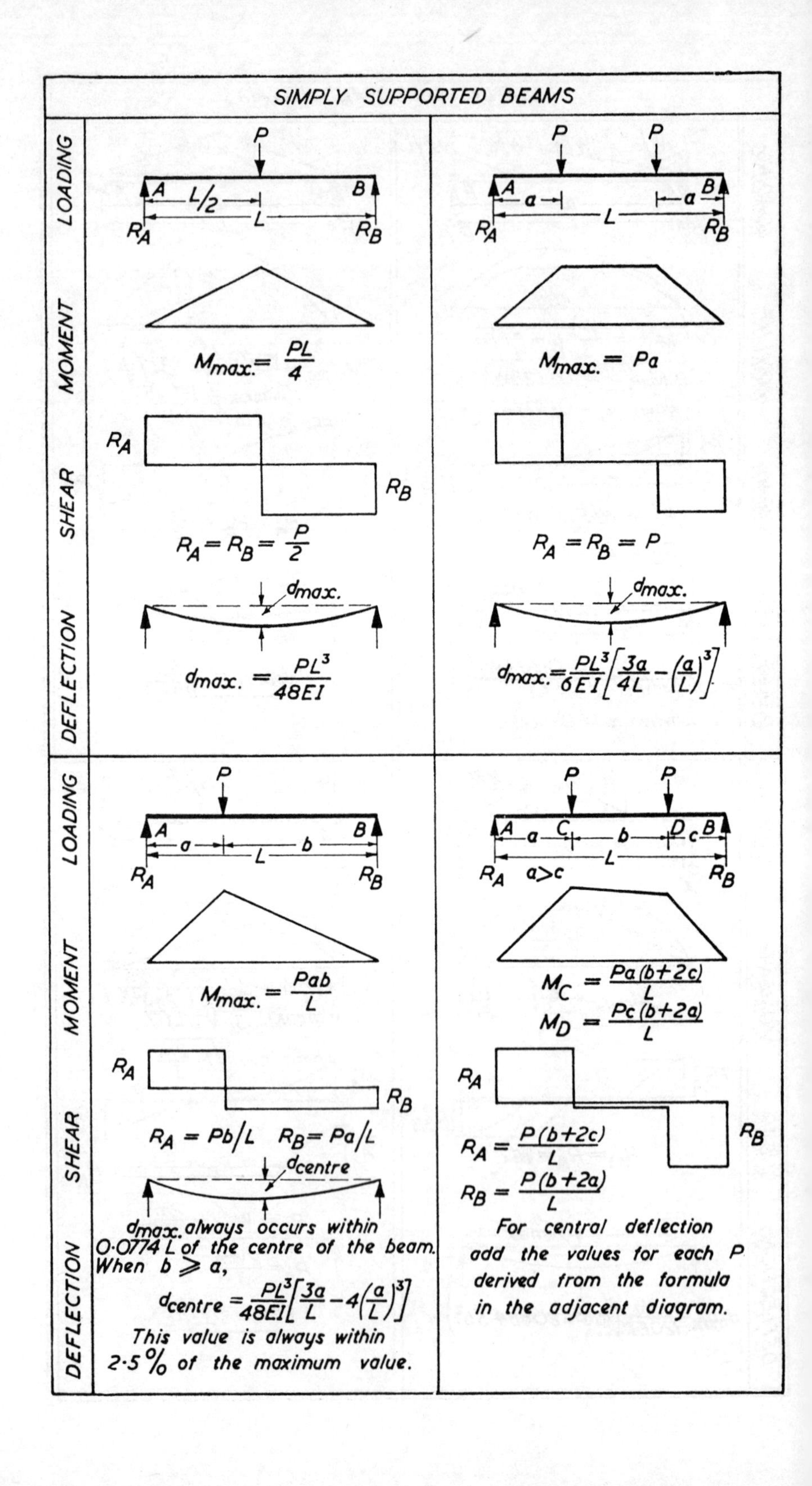
SIMPLY SUPPORTED BEAMS
LOADING
MOMENT
SHEAR
DEFLECTION
P
A
B
L/2
L
R_A
R_B
$M_{max.} = \frac{PL}{4}$
$R_A = R_B = \frac{P}{2}$
$d_{max.}$
$d_{max.} = \frac{PL^3}{48EI}$
P
P
a
a
L
$M_{max.} = Pa$
$R_A = R_B = P$
$d_{max.} = \frac{PL^3}{6EI}\left[\frac{3a}{4L} - \left(\frac{a}{L}\right)^3\right]$
a
b
$M_{max.} = \frac{Pab}{L}$
$R_A = Pb/L \quad R_B = Pa/L$
d_{centre}
$d_{max.}$ always occurs within 0·0774 L of the centre of the beam.
When $b \geqslant a$,
$d_{centre} = \frac{PL^3}{48EI}\left[\frac{3a}{L} - 4\left(\frac{a}{L}\right)^3\right]$
This value is always within 2·5% of the maximum value.
C
D
c
a>c
$M_C = \frac{Pa(b+2c)}{L}$
$M_D = \frac{Pc(b+2a)}{L}$
$R_A = \frac{P(b+2c)}{L}$
$R_B = \frac{P(b+2a)}{L}$
For central deflection add the values for each P derived from the formula in the adjacent diagram.

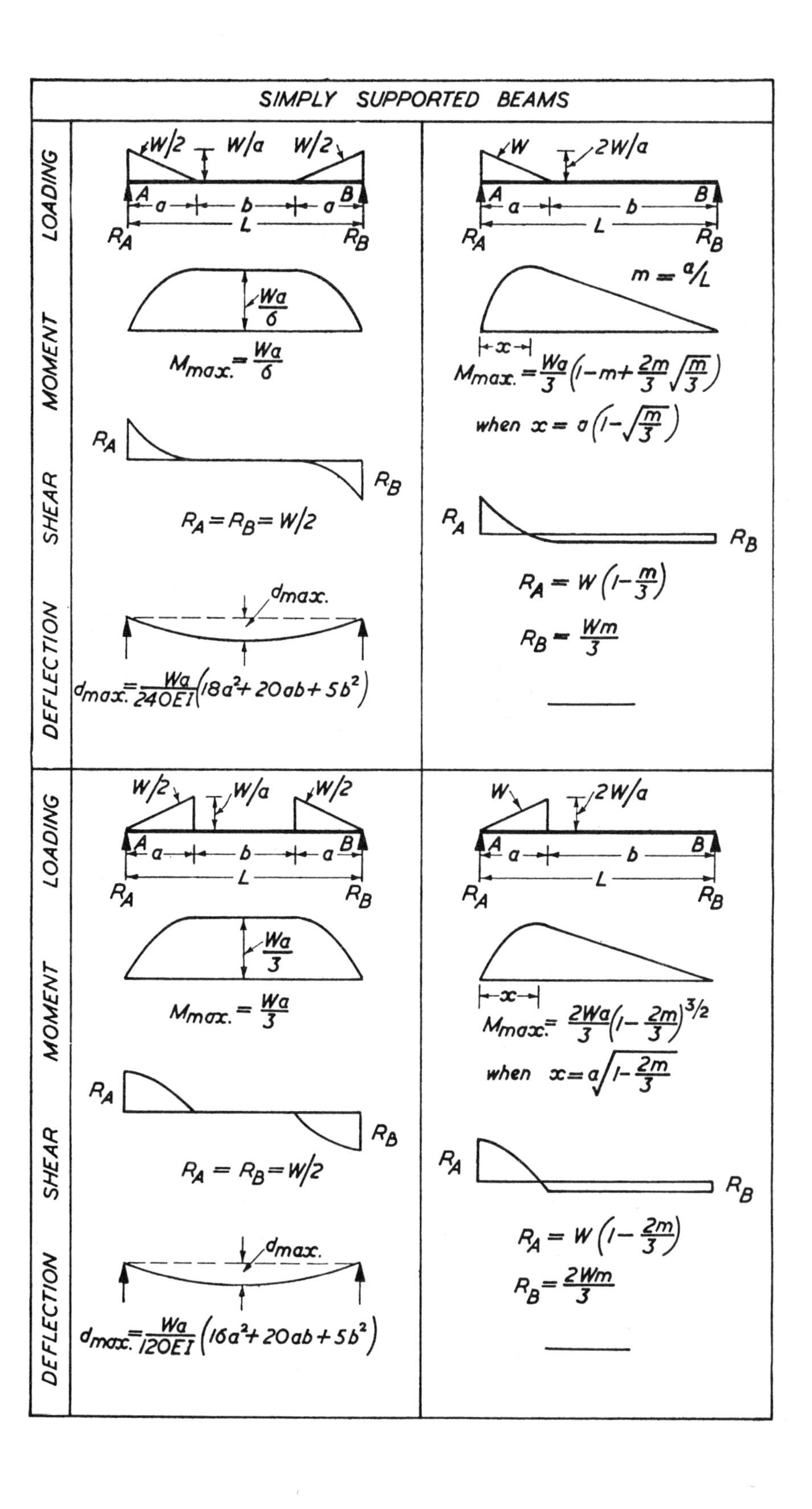

SIMPLY SUPPORTED BEAMS
LOADING
MOMENT
SHEAR
DEFLECTION
W/2
W/a
W/2
A
B
a
b
a
L
R_A
R_B
$\frac{Wa}{6}$
$M_{max.} = \frac{Wa}{6}$
R_A
R_B
$R_A = R_B = W/2$
$d_{max.}$
$d_{max.} = \frac{Wa}{240EI}(18a^2 + 20ab + 5b^2)$
W
2W/a
A
B
a
b
L
R_A
R_B
$m = a/L$
x
$M_{max.} = \frac{Wa}{3}\left(1 - m + \frac{2m}{3}\sqrt{\frac{m}{3}}\right)$
when $x = a\left(1 - \sqrt{\frac{m}{3}}\right)$
R_A
R_B
$R_A = W\left(1 - \frac{m}{3}\right)$
$R_B = \frac{Wm}{3}$
W/2
W/a
W/2
A
B
a
b
a
L
R_A
R_B
$\frac{Wa}{3}$
$M_{max.} = \frac{Wa}{3}$
R_A
R_B
$R_A = R_B = W/2$
$d_{max.}$
$d_{max.} = \frac{Wa}{120EI}(16a^2 + 20ab + 5b^2)$
W
2W/a
A
B
a
b
L
R_A
R_B
x
$M_{max.} = \frac{2Wa}{3}\left(1 - \frac{2m}{3}\right)^{3/2}$
when $x = a\sqrt{1 - \frac{2m}{3}}$
R_A
R_B
$R_A = W\left(1 - \frac{2m}{3}\right)$
$R_B = \frac{2Wm}{3}$

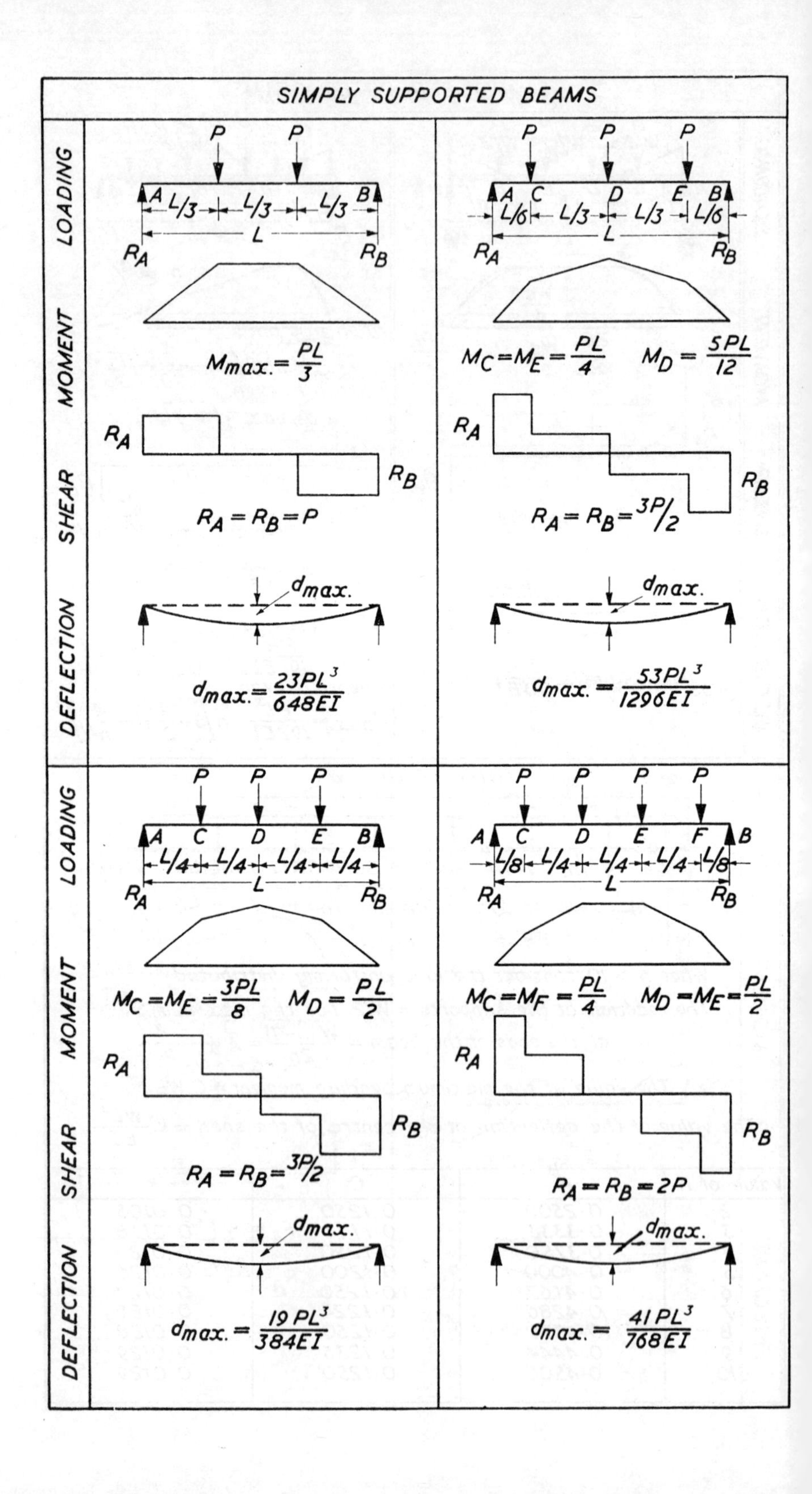
SIMPLY SUPPORTED BEAMS
LOADING
MOMENT
SHEAR
DEFLECTION
P P
A L/3 L/3 L/3 B
L
R_A R_B
M_max. = PL/3
R_A
R_B
R_A = R_B = P
d_max.
d_max. = 23PL^3/648EI
P P P
A C D E B
L/6 L/3 L/3 L/6
L
R_A R_B
M_C = M_E = PL/4 M_D = 5PL/12
R_A
R_B
R_A = R_B = 3P/2
d_max.
d_max. = 53PL^3/1296EI
P P P
A C D E B
L/4 L/4 L/4 L/4
L
R_A R_B
M_C = M_E = 3PL/8 M_D = PL/2
R_A
R_B
R_A = R_B = 3P/2
d_max.
d_max. = 19PL^3/384EI
P P P P
A C D E F B
L/8 L/4 L/4 L/4 L/8
L
R_A R_B
M_C = M_F = PL/4 M_D = M_E = PL/2
R_A
R_B
R_A = R_B = 2P
d_max.
d_max. = 41PL^3/768EI

SIMPLY SUPPORTED BEAMS

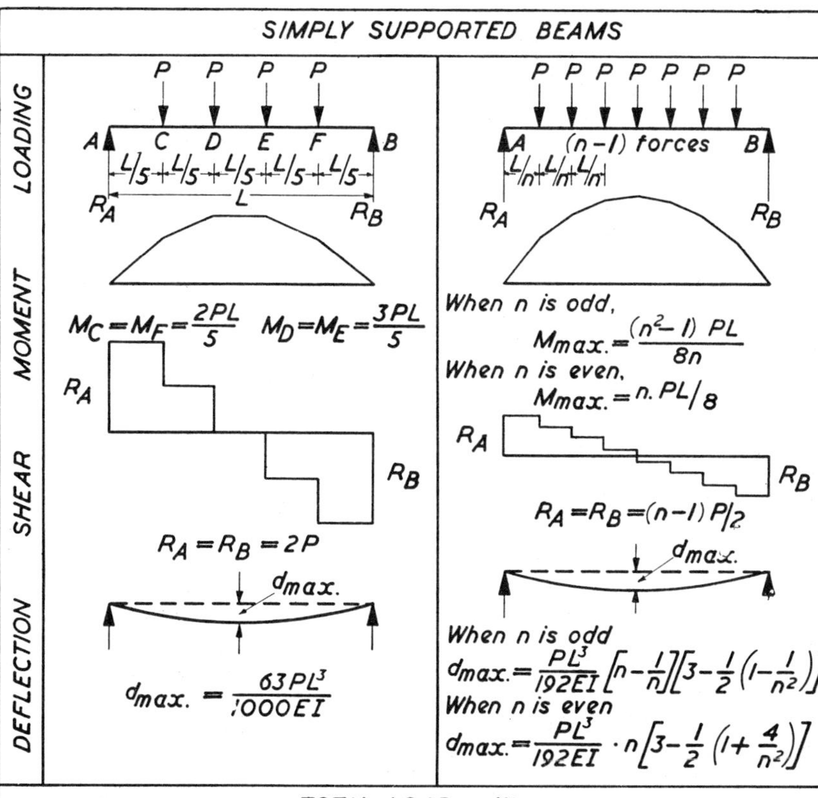

TOTAL LOAD = W

W/2n W/n W/n SIMPLY SUPPORTED BEAM W/2n

L/n L/n L/n L/n L/n L/n L/n

L

When $n > 10$, consider the load uniformly distributed

The reaction at the supports = $W/2$, but the maximum S.F. at the ends of the beam $= \frac{W(n-1)}{2n} = A.W$

The value of the maximum bending moment $= C.WL$

The value of the deflection at the centre of the span $= k.\frac{WL^3}{EI}$

Value of n	A	C	k
2	0·2500	0·1250	0·0105
3	0·3333	0·1111	0·0118
4	0·3750	0·1250	0·0124
5	0·4000	0·1200	0·0126
6	0·4167	0·1250	0·0127
7	0·4286	0·1224	0·0128
8	0·4375	0·1250	0·0128
9	0·4444	0·1236	0·0129
10	0·4500	0·1250	0·0129

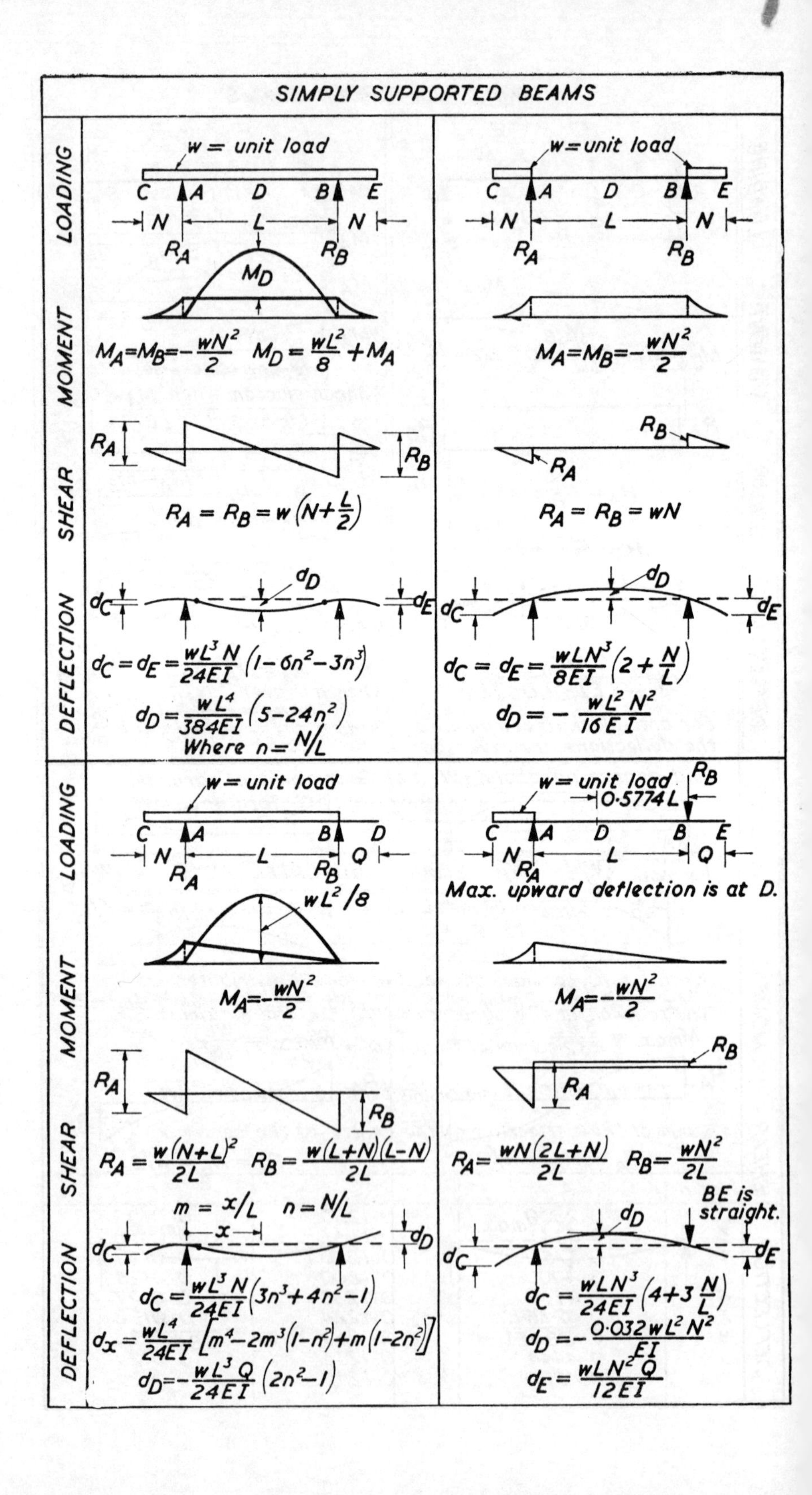

SIMPLY SUPPORTED BEAMS
LOADING
MOMENT
SHEAR
DEFLECTION
w = unit load
C A D B E
N L N
R_A R_B
M_D
$M_A = M_B = -\frac{wN^2}{2}$ $M_D = \frac{wL^2}{8} + M_A$
$R_A = R_B = w\left(N + \frac{L}{2}\right)$
d_C d_D d_E
$d_C = d_E = \frac{wL^3N}{24EI}(1 - 6n^2 - 3n^3)$
$d_D = \frac{wL^4}{384EI}(5 - 24n^2)$
Where $n = N/L$
w = unit load
C A D B E
N L N
R_A R_B
$M_A = M_B = -\frac{wN^2}{2}$
$R_A = R_B = wN$
d_C d_D d_E
$d_C = d_E = \frac{wLN^3}{8EI}\left(2 + \frac{N}{L}\right)$
$d_D = -\frac{wL^2N^2}{16EI}$
w = unit load
C A B D
N L Q
R_A R_B
$wL^2/8$
$M_A = -\frac{wN^2}{2}$
$R_A = \frac{w(N+L)^2}{2L}$ $R_B = \frac{w(L+N)(L-N)}{2L}$
$m = x/L$ $n = N/L$
d_C x d_D
$d_C = \frac{wL^3N}{24EI}(3n^3 + 4n^2 - 1)$
$d_x = \frac{wL^4}{24EI}\left[m^4 - 2m^3(1-n^2) + m(1-2n^2)\right]$
$d_D = -\frac{wL^3Q}{24EI}(2n^2 - 1)$
w = unit load
R_B
0·5774 L
C A D B E
N L Q
R_A
Max. upward deflection is at D.
$M_A = -\frac{wN^2}{2}$
R_B R_A
$R_A = \frac{wN(2L+N)}{2L}$ $R_B = \frac{wN^2}{2L}$
BE is straight.
d_C d_D d_E
$d_C = \frac{wLN^3}{24EI}\left(4 + 3\frac{N}{L}\right)$
$d_D = -\frac{0{\cdot}032\,wL^2N^2}{EI}$
$d_E = \frac{wLN^2Q}{12EI}$

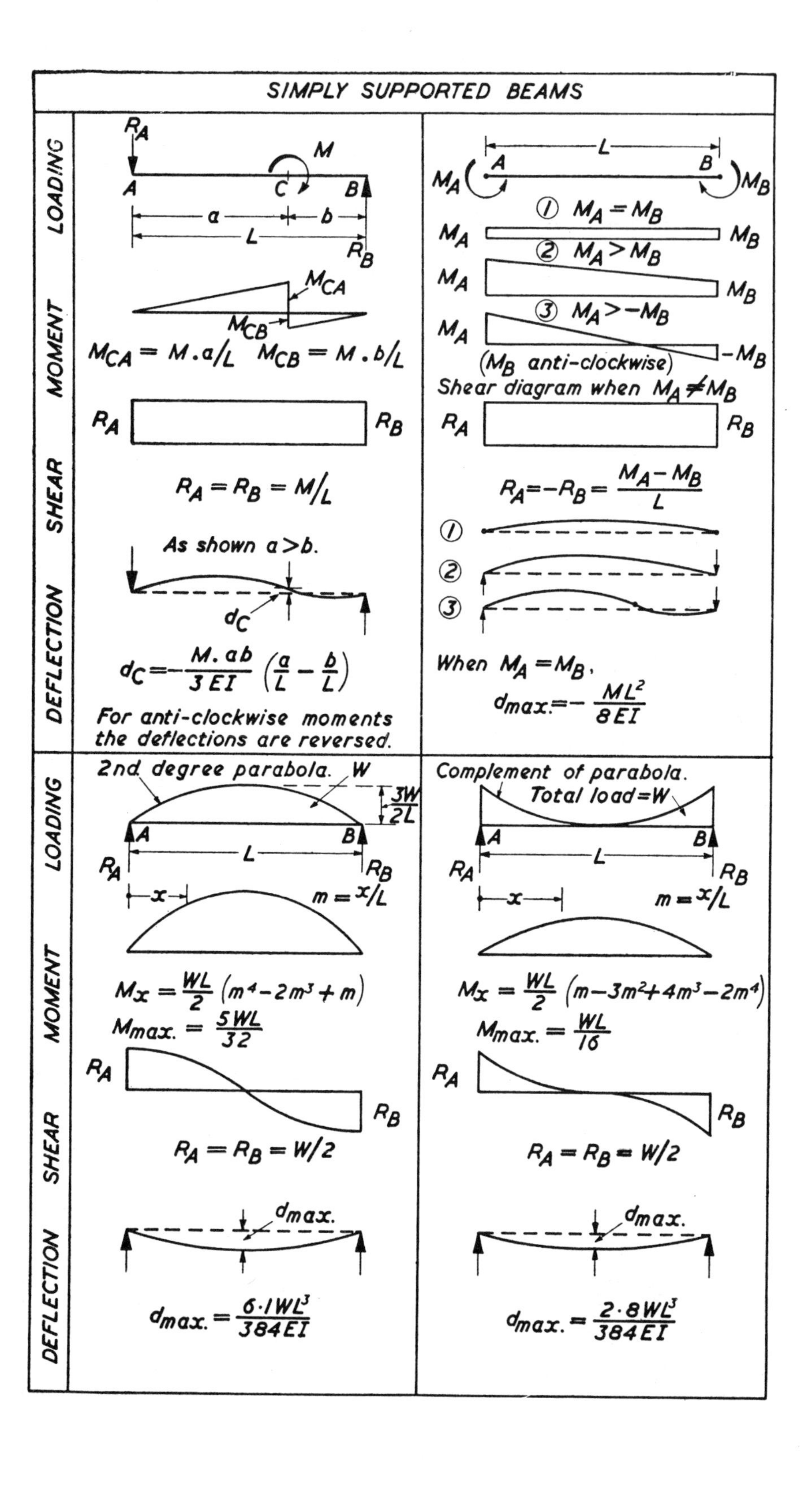

SIMPLY SUPPORTED BEAMS
LOADING
MOMENT
SHEAR
DEFLECTION
R_A
M
A
C
B
a
b
L
R_B
M_{CA}
M_{CB}
$M_{CA} = M . a/L$ $M_{CB} = M . b/L$
$R_A = R_B = M/L$
As shown a>b.
d_C
$d_C = -\frac{M.ab}{3EI}\left(\frac{a}{L} - \frac{b}{L}\right)$
For anti-clockwise moments the deflections are reversed.
M_A
M_B
① $M_A = M_B$
② $M_A > M_B$
③ $M_A > -M_B$
$-M_B$
(M_B anti-clockwise)
Shear diagram when $M_A \neq M_B$
$R_A = -R_B = \frac{M_A - M_B}{L}$
When $M_A = M_B$,
$d_{max.} = -\frac{ML^2}{8EI}$
2nd degree parabola.
W
$\frac{3W}{2L}$
x
$m = x/L$
$M_x = \frac{WL}{2}(m^4 - 2m^3 + m)$
$M_{max.} = \frac{5WL}{32}$
$R_A = R_B = W/2$
$d_{max.}$
$d_{max.} = \frac{6 \cdot 1WL^3}{384EI}$
Complement of parabola.
Total load=W
$M_x = \frac{WL}{2}(m - 3m^2 + 4m^3 - 2m^4)$
$M_{max.} = \frac{WL}{16}$
$d_{max.} = \frac{2 \cdot 8WL^3}{384EI}$

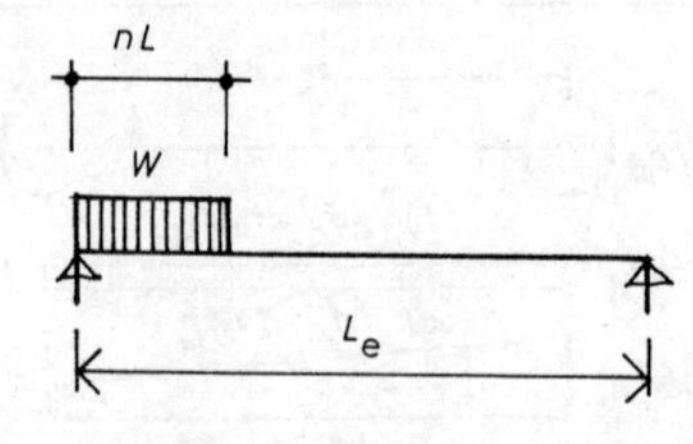

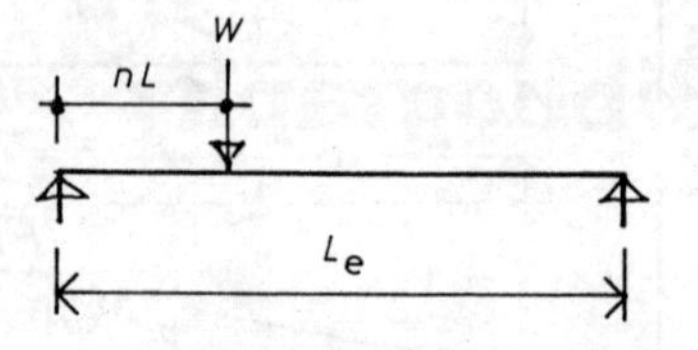

n	K_b
0.1	0.238
0.2	0.467
0.3	0.677
0.4	0.858
0.5	1.000
0.6	1.095
0.7	1.138
0.8	1.133
0.9	1.080
1.0	1.000

n	K_b
0.05	0.240
0.10	0.474
0.15	0.698
0.20	0.910
0.25	1.100
0.30	1.267
0.35	1.406
0.40	1.510
0.45	1.577
0.50	1.600

Bibliography

Materials

TRADA (1981) *Plywood – Its Manufacture and Uses*, TRADA (revised).
FPRL (1956) *A Handbook of Hardwoods*, HMSO, London.
FPRL (1957) *A Handbook of Softwoods*, HMSO, London.
FPDA *Finnish Birch Plywood and Blockboard*, FPDA.
CPA (1982) *Technical Manual 1982*, CPA.
COFI *CLS Kiln-Dried Spruce-Pine-Fir*, COFI, London.

Physical properties

Lavers, G. M. (1967) *The Strength Properties of Timber*, FPRL, HMSO, London.
Hudson, W. M. (1967) *The Strength Properties of European Redwood and Whitewood*, FPRL, HMSO, London.

Stress grading

Curry, W. T. (1966) *Grade Stresses for Structurally Laminated Timber*, FPRL, HMSO, London.
Lavers, G. M. (1967) *The Strength Properties of Timber*, FPRL, HMSO, London.
Sunley, J. G. (1968) *Grade Stresses for Structural Timber*, FPRL, HMSO, London.
Wilson, T. R. C. (1932) *Strength–Moisture Relations for Wood*, Technical Bulletin, No. 282, US Department of Agriculture.
British Standards Institution (1973) *Specification for Timber Grades for Structural Use*, BS 4978, British Standards Institution, London.
British Standards Institution (1971) *The Structural Use of Timber (Appendix A)*, CP 112, Part 2, British Standards Institution, London.

Durability and treatment

BRE Digest 72. *Home Grown Softwoods for Building*.
British Standards Institution (1977) *Preservative Treatments for Constructional Timber*, BS 5268, Part 5, British Standards Institution, London.
TRADA (1975) *Timber Preservation* (a handbook), BWPA, TRADA.

Regulations

Department of the Environment (1976) *The Building Regulations*, HMSO, London.
HMSO (1971) *The Building Standards (Scotland), (Consolidation) Regulations*, HMSO, London.

Design

British Standards Institution (1964) *Schedule of Weights of Building Materials*, BS 648, British Standards Institution, London.
British Standards Institution (1978) *Fire Resistance of Timber Structures* BS 5268, Part 4, Section 4.1, British Standards Institution, London.
British Standards Institution (1976) *Finger Joints in Structural Softwood*, BS 5291, British Standards Institution, London.
British Standards Institution (1971) *The Structural Use of Timber*, CP 112, Part 2, British Standards Institution, London.
British Standards Institution (1973) *Roof Trusses in Domestic Dwellings*, CP 112, Part 3, British Standards Institution, London.
British Standards Institution (1967) *Loading – Dead and Imposed*, CP 3, Chapter V, Part 1, British Standards Institution, London.
British Standards Institution (1972) *Wind Loads*, CP 3, Chapter V, Part 2, British Standards Institution, London.
Burgess, H. J. (1976) *Derivation of The Wall Racking Formulae in TRADA's Design Guide for Timber Frame Housing*, TRADA, London.
Coultas, H. W. (1961) *Theory of Structures*, Pitman, London (Chapter XIV, Flitch beams).
Morris, E. N. and Gajjar, S. (1981) Load-displacement relationship for nailed joints with solid timber and plywood members. *J. Inst. Wood Science*, 9(2), 62–64.
Newberry, C. W. and Eaton, K. J. (1974) *Wind Loading Handbook*, BRE, HMSO, London.
Ozelton, E. C. and Baird, J. A. (1976) *Timber Designers' Manual*, Crosby Lockwood Staples, London.
Reece, P. O. (1982) The design of continuous members in timber trussed rafters with punched metal connector plates. *J. Inst. Struct. Eng.*, 60A(5), 155–165.
Roark, R. J. (1956) *Formulas for Stress and Strain*, 4th edn, McGraw-Hill, New York.
Robertson, R. A. and Griffiths, D. R. (1981) Factors affecting the racking resistance of timber framed panels. *J. Inst. Struct. Eng.*, 59B(4), 49–63.

Index